U0112674

福建的世界遗产

World Heritage Sites in Fujian

福建省文化和旅游厅 策划
资源开发处 组织编写

编著
林叶

翻译
[加]Cassidy Gong
谢旻怡

摄影
刘贤健

海峡出版发行集团 THE STRAITS PUBLISHING & DISTRIBUTING GROUP | 福建人民出版社 FUJIAN PEOPLE'S PUBLISHING HOUSE

图书在版编目（CIP）数据

泰宁丹霞：汉英对照 / 林叶编著；（加）龚娴
（Cassidy Gong）等译 . --福州：福建人民出版社，
2020. 5

（福建的世界遗产）

ISBN 978-7-211-08312-1

Ⅰ. ①泰… Ⅱ. ①林… ②龚… Ⅲ. ①丹霞地貌—介
绍—泰宁县—汉、英 Ⅳ. ①P931

中国版本图书馆 CIP 数据核字（2019）第 288387 号

泰宁丹霞

TAINING DANXIA

作　　者：林　叶

翻　　译：［加］Cassidy Gong　谢旻怡

责任编辑：李文淑　林凤来

美术编辑：陈培亮

装帧设计：［澳］Harry Wang

内文排版：良之文化传媒

出版发行：福建人民出版社

电　　话：0591-87533169(发行部)

网　　址：http://www. fjpph. com

电子邮箱：fjpph7211@126. com

地　　址：福州市东水路 76 号

邮政编码：350001

经　　销：福建新华发行（集团）有限责任公司

印　　刷：雅昌文化（集团）有限公司

地　　址：深圳市南山区深云路 19 号

开　　本：787 毫米×1092 毫米　1/16

印　　张：15

字　　数：333 千字

版　　次：2020 年 5 月第 1 版

印　　次：2020 年 5 月第 1 次印刷

书　　号：ISBN 978-7-211-08312-1

定　　价：88. 00 元

目 录

Contents

01

世遗档案

UNESCO's Introduction to China Danxia

◎ 大金湖情侣峰。

Lovers' Peak by Dajin Lake.

名　　称：中国丹霞

列入时间：2010年8月被正式列入《世界遗产名录》

遗产类型：世界自然遗产

中国丹霞是中国境内由陆相红色砂砾岩在内生力量（包括隆起）和外来力量（包括风化和侵蚀）共同作用下形成的各种地貌景观的总称。这一遗产包括中国西南部亚热带地区的六处遗产。它们的共同特点是壮观的赤壁丹崖以及一系列侵蚀地貌，包括雄伟的天然岩柱、岩塔、沟壑、峡谷和瀑布等。这里跌宕起伏的地貌，对保护亚热带常绿阔叶林和包括约400种珍稀或濒危物种的动植物起到了重要作用。

◎ 大金湖大赤壁。
Dachibi (the Grand Red Cliffs) by Dajin Lake.

China Danxia

Inscribed in 2010 on the World Heritage List

Naltural Property

China Danxia is the name given in China to landscapes developed on continental red terrigenous sedimentary beds influenced by endogenous forces (including uplift) and exogenous forces (including weathering and erosion). The inscribed site comprises six areas found in the sub-tropical zone of southwestern and southeastern China. They are characterized by spectacular red cliffs and a range of erosional landforms, including dramatic natural pillars, towers, ravines, valleys and waterfalls. These rugged landscapes have helped to conserve sub-tropical broad-leaved evergreen forests, and host many species of flora and fauna, about 400 of which are considered rare or threatened.

遗产价值

中国丹霞是一个由赤水、泰宁、崀山、丹霞山、龙虎山和江郎山六个部分组成的系列遗产，位于中国偏东南部亚热带地区，自贵州省由西向东至浙江省约1700千米的新月形弧内。

- 丹霞地貌从新近纪至今一直以红色沉积序列发育。六个部分代表了丹霞地貌从“最小侵蚀”到“最大侵蚀”的最重要例子，提供了丹霞地貌现象的一系列不同方面，并说明了地貌景观与形成它们的力和过程之间的关系。

- 中国丹霞是一个令人震撼的独特景观，极具自然之美。红色的砾岩和砂岩形成壮观的山峰、石柱、石壁和峡谷，与茂密的森林、蜿蜒的河流、雄伟的瀑布相呼应，呈现出绝妙的自然景观。

- 中国丹霞具有极高的完整性，境内有山峰、石塔、石台、石柱、石壁、峡谷、岩洞、岩拱等多种发育良好的丹霞特征地貌，反映了温带季风气候条件下大陆性地壳发育的特定阶段，呈现出从“年轻”到“成熟”再到“老年”的清晰地貌序列，每个阶段的地貌特征明晰。

- 中国丹霞包含足够规模的实体元素，反映了丹霞地貌从“年轻”到“成熟”再到“老年”各个阶段的自然美学价值和地球科学价值，缓冲地带也有明显界定。

◎ 坐竹筏游上清溪。
Drafting down Shangqing Stream.

Outstanding Universal Value

China Danxia is a serial property comprising six component parts (Chishui, Taining, Langshan, Danxiashan, Longhushan, and Jianglangshan) found in the sub-tropical zone of southeastern China within approximately 1700 km crescent shaped arc from Guizhou Province in the west to Zhejiang Province in the east.

◆ The Danxia landforms have been developed in red sedimentary sequences continuously from the Neogene until the present. The six component parts represent the most important examples of "least eroded" to "most eroded" Danxia landforms, providing a range of different aspects of the phenomenon, and illustrate both the range of landforms in relation to the forces and processes that formed them, together with a range of associated landscapes.

◆ China Danxia is an impressive and unique landscape of great natural beauty. The reddish conglomerate and sandstone that form this landscape of exceptional natural beauty have been shaped into spectacular peaks, pillars, cliffs and imposing gorges. Together with the contrasting forests, winding rivers and majestic waterfalls, China Danxia presents a significant natural phenomenon.

◆ China Danxia satisfies the requirements of integrity. The property contains a wide variety of well developed red-beds landforms such as peaks, towers, mesas, cuestas, cliffs, valleys, caves and arches. Being shaped by both endogenous forces (including uplift) and exogenous forces (including weathering and erosion), China Danxia provides a range of different aspects of the phenomenon of physical landscape developed from continental (terrestrial) reddish conglomerate and sandstone in a warm, humid monsoon climate, illustrating both the range of landforms in relation to the forces and processes that formed them. The component parts represent the best examples of "least eroded" to "most eroded" Danxia landforms, displaying a clear landform sequence from "young" through "mature" to "old age", and with each component site displaying characteristic geomorphologic features of a given stage.

◆ China Danxia encompasses substantial elements of sufficient size to reflect the natural beauty and earth science values of Danxia landform from young stage through mature stage and to old stage. The boundaries of the China Danxia are adequate in relation to the nominated earth science and aesthetic values, and the buffer zone boundaries are also clearly defined.

◎ 秋日泰宁水韵。

Beautiful water scene in autumn.

UNESCO
United Nations
Educational, Scientific and
Cultural Organization
World Heritage
Convention
保护世界文化
与自然遗产公约
世界遗产委员会
已将
中国丹霞
列入世界遗产名录
列入该名录是对一项为了
全人类利益而应加以
保护的文化或自然遗产的
突出普遍价值的确认
列入日期
2010年8月2日
伊琳娜 博科娃
联合国教科文组织总干事
中国丹霞·福建泰宁

UNESCO
United Nations
Educational, Scientific and
Cultural Organization
World Heritage
Convention
CONVENTION CONCERNING
THE PROTECTION OF
THE WORLD CULTURAL
AND NATURAL HERITAGE
The World Heritage Committee
has inscribed
China Danxia
on the World Heritage List
Inscription on this List confirms the outstanding
universal value of a cultural or
natural property which requires protection for
the benefit of all humanity
DATE OF INSCRIPTION
2 August 2010
DIRECTOR-GENERAL
OF UNESCO
Fujian Taining, China Danxia

02

走进泰宁

Tours around Taining

◎ 碧水丹山美如画。
Amazing red Danxia mountains and clear lake.

听音知韵——打开泰宁，从声音开始

到一个地方，可以用眼睛流连风景，用味蕾感受美食，也可以先从耳朵开始，触碰、感知一方风情。

属于泰宁的声音是什么？

首先是流水、虫鸣、竹响、山风，这些泰宁的天籁之声，四季皆动人；还有巷口吆喝，擂茶研磨，古井打水，小城春秋的烟火生活；还有僧人诵经、梅林戏咿呀、大源傩舞之鼓咚咚回荡，这些传统之音流传千年，你完全可以先不用导游，不用旅游手册，只需竖起耳朵，静静地聆听，感受没有任何先入为主的泰宁风韵。所谓妙不可言就是如此。

◎ 泰宁大金湖。
Dajin Lake in Taining.

Sound and Rhythm:To Understand Taining, Beginning with Sound

When we arrive at a place, we may view its scenery with our eyes, or taste its delicious food with our taste buds. We can also begin with our ears, to feel and understand local customs with our ears.

So, what is the sound of Taining?

It is first Taining's beautiful sound of nature through all four seasons—the sound of water flow, insects, bamboos and mountain winds. It is the sound of the small-town life throughout the year, people shouting from the head of the lane, tea making and grinding, and the sound of villagers getting water from the ancient wells. There is also the sound of monks chanting, Meilin Play, and the drum beats of the Nuo Dance in Dayuan—these are sounds of traditions that have been passed on for more than a thousand years. You may first visit Taining without a tour guide or a travel book; you can feel the charm of Taining without any biased opinion, by using your ears to listen carefully and quietly. This will be such a wonderful experience for you.

泰宁人的腔调

方言，是打开泰宁的第一把声音“钥匙”。

福建省历来以方言复杂著称。《泰宁县志・方言》开篇即写道：“泰宁县古代属邵武军、邵武府。方言原来也是闽北方言……由于历史和地理的原因，泰宁县的方言与邻县有许多差异。”

造成这个现象的原因之一是地理阻隔，“一直到新中国成立前，泰宁县还没有通汽车，对外交流主要靠杉溪小木船和金溪上的中型木船与下游的将乐、顺昌、南平联系，泰宁话较之其他三县保留了更多的闽方言的特点。然而在充满险滩的航道上，船运能力毕竟有限，在长期的自然经济为主体的社会生活中，大多数人极少出门，和外界交往不多的情况使泰宁话成了一种和周围各县还不甚相近的独特的方言……”

另一方面，“在数百年间，由于这里和赣东的抚州、南城、黎川等地往来密切，不少江西人来此经营手工业、商业，并陆续在此定居，这里的方言就渗透了许多赣方言的特点，形成了一个以闽方言为老底的赣语化的方言区。”

The Tone of Taining People

The dialect is the first "key" of sound to open the door of Taining.

Fujian Province has always been known for its complicated dialects. *Taining County Records • Dialect* begins with, "Taining County belonged to Shaowu prefectural government in ancient times. The dialect was originally Northern *Min* (*Min* is short for Fujian Province) dialect... Due to the historical and geographical reasons, there are many differences between the dialects of Taining and the dialects of neighboring counties."

Geographical barrier is one reason for this. "Up until the founding of the People's Republic of China, there were no roads for cars to access to Taining County. Communication with the outside mainly depended on small wooden boats on Shanxi Stream and medium-sized wooden boats on Jinxi Steam with nearby counties including like Jiangle, Shunchang and Nanping. As a result, Taining dialect kept more characteristics of *Min* dialect than the other three dialects. However, in the channels full of dangerous shoals, the shipping capacity is limited after all. In addition, in a society where natural economy dominates for a long time, most people rarely go out. The lack of contact with the outside world has made Taining dialect a unique dialect that is not similar to any surrounding counties..."

On the other hand, "for hundreds of years, because of the close relationship between Taining and cities in the east of Jiangxi Province like Fuzhou, Nancheng, Lichuan, many people from Jiangxi came here to engage in business in handicraft, commerce and gradually settled here permanently. As a result, the dialect here has infiltrated many characteristics of *Gan* dialect (*Gan* is short for Jiangxi Province), forming a dialect area based on *Min* dialect with gradual influence of *Gan* dialect."

◎ 打锡壶。
Making a tea pot with tin.

“打锡壶，补锅咯——”这是已经难得听到的匠人们的沿街吆喝。“汝几若好，麻烦汝（你真好，麻烦你了）”，这是乡里间客气的对答。走在泰宁的街头村尾，难懂方言，却感受到一份地道。如果你还能偶然听到些原汁原味的用本地方言演唱的山歌、民谣、小调之类的，绝对是万分之幸。

"Da—xi—hu, bu—guo—lo—" ("Making tea pots with tin, mending pots"). This is the hawk of the craftsmen walking down the street but can rarely be heard anymore. "Ru ji ruo hao, ma fan ru" ("You are so kind, thanks for the trouble"), this is a polite greeting of the locals in the township. Though it is hard to understand the dialect of Taining, you can always feel the kindness and authenticity when walking on its streets. You are considered extremely fortunate if you all of sudden hear some original singing in the local dialect of folk songs, ballads or minor tunes.

泰宁城小，过去一直流传着这样一首民谣：“小小泰宁县，三家豆腐店，城内磨石腐，城外听得见。”

泰宁梅林戏俗称土京戏，最流行的时候人人都会唱几段。“梅林十八坊，十个弟子九担箱；敲起叮咚鼓，唱起摩郎腔。茅担抬石臼，抬到塅中央；搭起土台子，唱到大天光。”

Taining Town is small, with a folk song spreading in the past, "Taining Town is so small that when the three tofu shops are grinding tofu with millstone inside, one can hear the sound from outside the town."

Meilin Play of Taining is known as the local Peking Opera. Everyone could sing a few lyrics during its peak period of popularity. "The Shibafang Meilin Opera Troupe, with ten disciples, nine boxes. They beat the drums and sing with the Molang Tone. They use carrying poles to move stone mortar to the center of an open flat and set up the stage, then perform from night to the next morning."

◎ 梅林村十八坊民间艺术团是当地非常受欢迎的剧团。
The Shibafang Meilin Opera Troupe of Meilin Village is very popular in the area.

◎ 表演泰宁山歌小调《砍柴歌》。
Performance of *Chopping Wood Song*.

泰宁盛产笋竹，笋竹可加工成“闽西八大干”里的笋干。因此泰宁自古就流传着不少和笋有关的山歌小调，现在在新桥乡一带还有传唱。比如《谷雨以后再回家》：“少女妻子一枝花，人在笋厂心在家。家中父母年又老，谷雨以后再回家。”还有《砍柴歌》：“刀斧利锋强，劈樵好几方。一来是煮饭，二来煮笋要。若要做好笋，劝君早备樵。莫等做笋时，又愁没柴烧。”

Taining is rich in bamboo shoots, which can be processed into dried bamboo shoots—one of "the Eight Famous Dried Food of Fujian". Since ancient times, many folk songs and ditties related to bamboo shoots have been passed on in Taining and are still sung in Xinqiao Township. For example, *Go Home after Grain Rain,* "Though the young men are working at the bamboo factory, their hearts are at home because of their flower-like young wives. Their parents are old, and they go home after the Grain Rain." There is also *Chopping Wood Song,* "Axes that are sharp are the best for cutting much firewood. The firewood is used not only for cooking rice but also for cooking bamboo shoots. To make good dried-bamboo shoots, you are suggested to prepare enough firewood in advance. Don't wait until you start to cook the bamboo shoots and find there is no firewood."

◎ 表演唱泰宁山歌。
Singing mountain songs.

做笋干之前要先煮笋，所以要提前准备好木柴，山上砍的木柴是湿的，要先自然晾干，不然到时候就没法顺利燃烧。道理简单通俗，却也蕴含了质朴的生活哲学。俗话说："唱戏一半假，山歌句句真。"山歌是自发的，自由的，也是最地道的。

"八月十五怎团圆？世上道路最唔平，天浪星多月不亮，人间官多理唔明。""茶树花开连打连，今年开花果明年。艄公撑船等大水，十八妹子等少郎。"上清溪和九龙潭上的艄公，兴起也会唱几首山歌给游客听："山里的歌儿，实在多哟。山里的歌儿，实在美呀。山里的歌儿，真好听咯。山里的人儿，喜欢唱山歌哟。"

To make dried bamboo shoots, it is important to prepare dried wood before cooking the bamboo shoots. The firewood that are freshly cut from up in the mountains are wet, so they must be dried first or they cannot burn smoothly. It is a simple logic yet contains philosophy of life. As the saying goes, "Dramas are half made-up, but there is some truth in every line of the folk songs." Indeed, folk songs are spontaneous, free and authentic.

"How to reunite in the Mid-Autumn Festival? The way of life is the least smooth; the moon cannot shine when there are too many stars in the sea sky; reasons cannot be clarified when there are too many government officials." "The tea trees are in full blossom and these blooms will turn into fruit next year. The boatmen punt the boat waiting for water, like eighteen-year-old girls wait for their young lovers." The boatmen of Shangqing Stream and Jiulong (Nine-Dragon) Pond will sing a few folk songs to the tourists when they are in the mood, "The mountain songs are so plenty, the mountain songs are so beautiful, the mountain songs are so pleasant to hear, and mountain people like singing mountain songs."

上青乡还有种叠罗汉的民俗民艺。一组六人，其中三人相叠，上面人的脚要踩在下面人的肩上，像杂技表演。最上面那人拿着锣，中间那人拿着鼓，最下面那人拿着戒尺。边上站着另外三人，站在正面的那人拿着木鱼，站在边上的两人一人拿鼓，一人拿锣。叠好后，大家边敲边唱。唱词叫《罗汉歌》：“……九下鼓，九下锣，多情小姐听我来唱歌。多情小姐听我唱：选个勤劳好丈夫。十下鼓，十下锣，男女老少听我来唱歌。男女老少听我唱，打鼓打锣闹得笑呵呵。”

和很多地方一样，普通话的推广，交通的便捷，城市化的进程，让泰宁古老的音韵式微，但它们才是泰宁的灵魂。

There is also a custom folk art at Shangqing Township called Dieluohan, which means piling up the arhats. During the performance, six people are divided into two groups of three, one group forms a human pillar of three, where the feet of the man above must step on the shoulders of the man below just like an acrobatic show. The person at the top holds a gong, the one in the middle holds a drum, and the one at the bottom holds a ferule. Other three men standing around, the one in the front holding a wooden fish, and the two people on both sides holding a drum and a gong respectively. When they finish "piling up". They sing while beating the gong and the drum. The lyrics of the *Arhats Song* goes, "... nine drum-beats, nine gong-beats, amorous lady please listen to me sing. Please listen to me: to find a diligent husband..."

Like many places, the standardization of Mandarin, the convenience of transportation and the process of urbanization have weakened the ancient sounds of Taining despite that they are the soul of Taining.

梅林戏，搭起戏台唱到大天光

梅林戏又叫土京戏，是徽调班经浙江、江西传入福建后形成的一种乡土戏，因发祥于泰宁县朱口镇梅林村而得名。它被总结为“由经过江浙赣入闽的中原文化（徽调）和闽越部落上古文化（山歌、巫术、道教音乐）交汇碰撞的火花”。

早在徽调传入之前，梅林村由于道教活动频繁，在当地已有道士腔的演唱活动。演绎至今的梅林戏源起的故事是：清乾隆年间，泰宁县朱口镇梅林村有一位寡妇周氏，家境富裕。为了祝寿，她请浙江的徽班到家里演戏，结果越看越上瘾，为了随时能听，干脆出钱请徽班的师傅传艺给当地的子弟。就这样，梅林戏落地生根，先是用本地的土官话唱，但在流传中难免走调，加上与本地的道教音乐、山歌小调等元素的不经意融汇，慢慢形成了一种独特的风格。

清末至民国初期是梅林戏的兴盛时期。早期梅林戏艺人半农半艺，农忙时务农，农闲时演戏，人称梅林戏班为“四季班”。全班有14人，角色分“三生、四旦、三花脸”，合称“十个子弟”，服装和化装相对简单，以徽调皮黄为主要声腔，演唱时生旦以大嗓起音，以小嗓落音，每个唱段尾声都有“伊子呀”的拖音，依稀保留山歌的余韵。梅林戏的表演中还有“变脸”“耍叉”“耍獠牙”“挺僵尸”“下高台”等传统特技。

1960年，泰宁县成立梅林戏剧团。2006年5月，泰宁梅林戏入选第一批国家级非物质文化遗产名录。现在要看梅林戏，可以到泰宁尚书第旁的泰宁梅林戏展示中心，这里每天上演《梅林谣》《看大戏》《背子赶会》等剧目的折子戏，表演夸张诙谐，让南腔北调的观众都看得很开心。

◎ 泰宁梅林戏剧团。

The Meilin Opera Troupe of Taining.

Meilin Opera—Building the Stage and Singing till the Next Morning

Meilin Opera, also known as Local Peking Opera, is a kind of local opera formed after Anhui Opera was introduced to Fujian through Zhejiang and Jiangxi provinces. It was named after its birthplace Meilin Village, Zhukou Town, Taining County. It is summed up as "the sparks of the interaction between the Central Plains culture (Anhui Opera) including cultures of Jiangxi, Zhejiang and Jiangxi before arriving in Fujian, and the ancient culture (folk songs, witchcraft, Taoism music) of the Minyue tribes".

Even before Anhui Opera was introduced, Meilin Village had already had Taoist-tone singing due to the frequent Taoist activities. The story of the origin of Meilin Opera was that during 1736—1795 in the Qing Dynasty, a wealthy widow Zhou of Meilin Village invited an Anhui Opera troupe from Zhejiang to perform on her birthday. The more she watched the performance, the more she liked it. In order to be able to watch an Anhui opera any time she wanted, Zhou funded local children to learn from artists of the Anhui Opera troupe. This was how Meilin Opera planted its seeds and grew in Meilin Village. Meilin Opera was first sung in local dialect, and then formed a unique style during the spread, because it would inevitably go out of tone, plus inadvertent integration of elements like local Taoism music and folk songs.

Meilin Opera flourished between the end of the Qing Dynasty and early years of the Republic of China. Early Meilin opera players were half artists half farmers—they performed the opera when farming was not busy. This was why people called Meilin Opera Troupe the "four-season troupe". The 14-person troupe included three Sheng (male roles), four Dan (female roles) and three Hualian (painted faces), which were altogether called "ten drama actors". Their costumes and makeup were relatively simple, and the main tunes are xipi and erhuang in Anhui Opera. When performing, the Sheng starts with true voice, ends with a falsetto voice, and each paragraph ends with a long sound of "Yi—zi—ya—", vaguely retaining the characteristics of folk songs of the mountains. Meilin Opera also performs traditional stunts such as changing face, playing the fork, playing fangs, and getting off the high platform.

The Meilin Opera Troupe of Taining was established in 1960. In May 2006, it was listed in the first group of national intangible cultural heritages. Nowadays, people can go to Meilin Opera Exhibition Center of Taining next to Shangshu (the Minister) Mansion to see Meilin operas. Famous traditional operas like *Meilin Ballad, Watching a Play, Carrying the Baby to the Temple Fair* are performed at the centre every day. The exaggerated and humorous performance make the audiences laugh all the time.

◎《梅林谣》。

Meilin Ballad.

◎《背子赶会》。

Carrying the Baby to the Temple Fair.

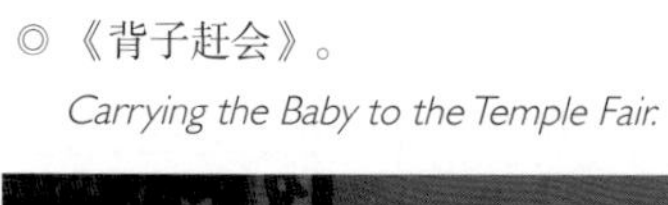

◎ 新编历史故事剧《花蕊夫人》。
Historical play *Madam Huarui.*

◎ 新编历史故事剧《画网巾》。
Historical play *Drawing Manggeon.*

◎ 新编历史故事剧《邹应龙》。
Historical play *Zou Yinglong.*

摩郎腔和上青古乐

“先拜杨公二拜龙，三拜祖师到地中。南极仙翁作师主，九天玄女定吉凶……”这是《安祀歌》，是道士在死者棺材下葬后，于墓碑前设坛做法时所唱。

泰宁人把道士叫作摩郎仙。过去在泰宁，人一生中的很多重要时刻，如生死、婚娶、乔迁、寿庆，都要请摩郎仙举行一些仪式来祈福禳灾，而在各种法事中，都有音乐紧密配合。摩郎仙头戴如意冠，身穿八卦双龙道袍，等鼓乐响起，道士开始高唱《开坛赞》《神农赞》《三宫颂》《开天门》《上清宝赞》等道曲，歌声激越悠扬，钟鼓节奏明快，会把人带入庄严肃穆的氛围之中。

原福建省文化厅公布了福建省第四批非物质文化遗产代表性项目代表性传承人名单，黎基求作为泰宁上青古乐的传承人入选，让泰宁上青古乐进入了更多人的视野。

◎ 泰宁上青古乐队演奏《上清宝赞》。
Performance of *Praise to God Shangqing* by Shangqing Ancient Music Orchestra.

◎ 泰宁上青古乐传承人黎基求。
Li Jiqiu, an inheritor of Shangqing Ancient Music.

Molang Tone and Shangqing Ancient Music

"First kowtow to Yang Gong, then kowtow to the dragon, lastly kowtow to the ancestors buried underground..." This is *Ansi Song* (the song of a sacrificial offering), sung by Taoist priests when they set up altars in front of tombstones after the dead are buried in coffins.

People in Taining call Taoist priests Molang celestial beings. In the past, Taining people would invite Molang to hold ceremonies at many important moments in their life including birth, death, marriage, housewarming and birthday celebrations, to pray for happiness and guard disasters from happening. All the rituals were accompanied by music. When the drum music started, the Taoist priest wearing a fortune crown and a double-dragon robe with the Eight Diagram imprinted, sung sutras such as *Praise to the Altar*, *Praise to Shennong* (legendary god of farming), *Praise to the Three Palaces*, *Open the Heaven Gate* and *Praise to God Shangqing*. The melodious singing and the lively rhythm of the bell and drum will bring people into a solemn atmosphere.

The Bureau of Culture in Fujian announced the fourth list of representative inheritors of intangible cultural heritage representative project of the province. Li Jiqiu was selected as the inheritor of Taining Shangqing Ancient Music, which made the music known by more people.

上青古乐，之前更多是被称为上青道乐，指的是泰宁一带的道士在举行法事活动时所运用的音乐，有独唱、齐唱、吟唱的歌曲和吹奏、弹拉、打击器乐演奏的曲牌，继承并发展了宋代以来流行于泰宁周边的清徽派和普庵派道教音乐以及南词、传统民歌和民间吹奏乐的精华。这其中，大部分唱曲和吹奏乐为泰宁和周边区域所独有，对研究泰宁本地民俗文化也有很高的价值。

道教音乐继承了中国古代宫廷音乐和传统民间音乐的基因，是中国传统音乐的一脉，又称法事音乐、道场音乐，大约开始于南北朝时期。唐代是道教音乐发展的鼎盛时期之一，唐玄宗曾命道士、大臣献道曲，并亲自研作和教授道乐，用于在太清宫祭献老子时演奏。宋太宗、宋真宗、宋徽宗分别编写道教音乐，多达数十首，如《步虚辞》《散花词》《白鹤赞》《玉清乐》《太清乐》等。宋徽宗还曾经举国选宫观道士进京习道乐。明太祖朱元璋设玄教院统辖全国道教。

道教在泰宁发展的历史很悠久。早在西汉末年，江西南昌的尉梅福就在上清溪结庐炼丹，有当地村民拜其为师，并代代相传，道教音乐也随之在泰宁一直流传至今。但随着社会变迁，老一辈艺人相继离世，上青道乐濒危。据黎基求回忆，在其祖上留下的资料中，关于道教音乐的原有十几本，其中工尺谱演唱和演奏曲目、曲牌达2000多首，但目前就剩下200多首。

◎《工尺谱》（上青道教曲）原稿。

Gongchi Score (a traditional Chinese musical notation).

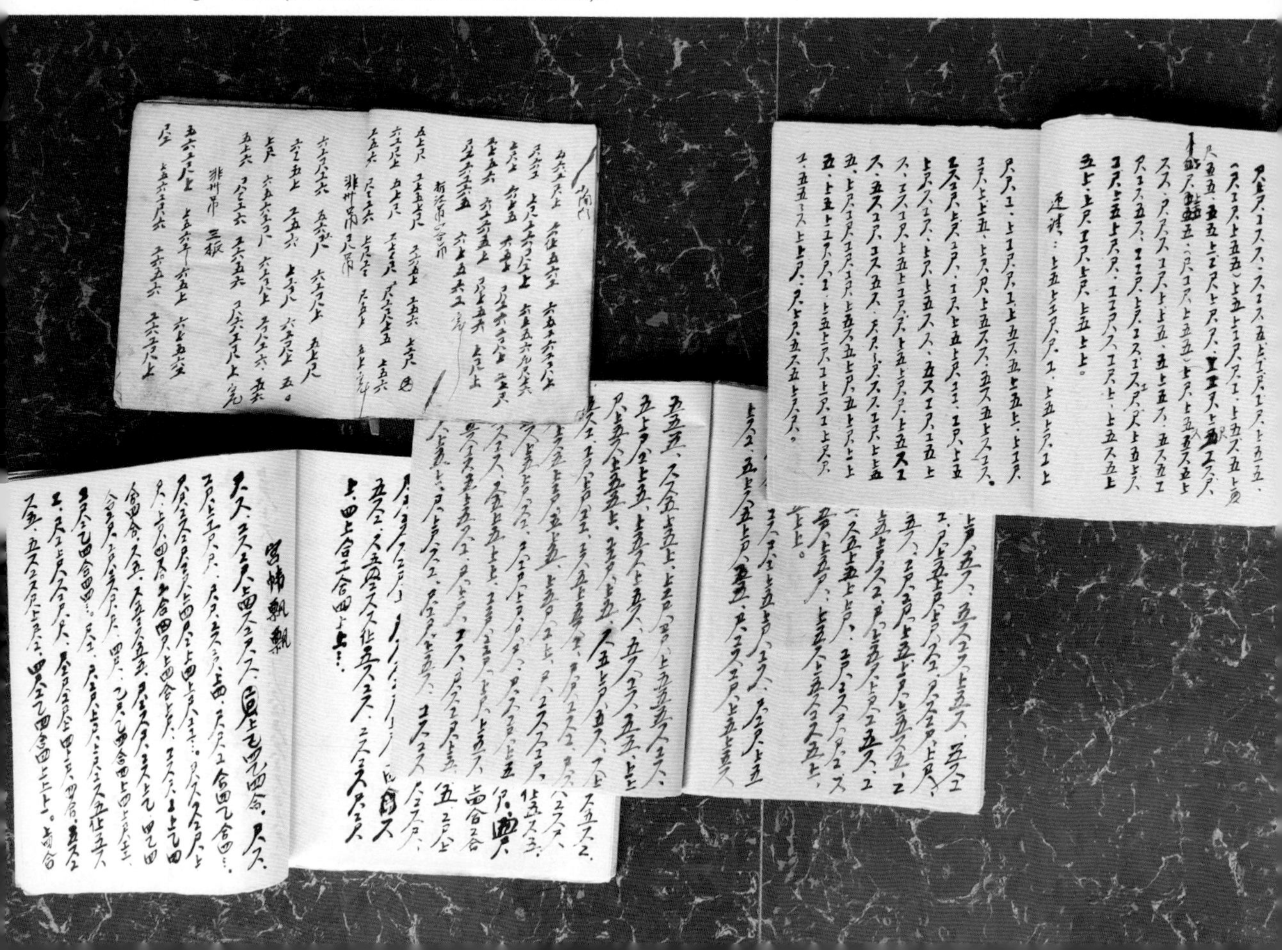

Shangqing Ancient Music, formerly known as Shangqing Taoist Music, refers to the music used by Taoist priests in Taining area during ritual activities. There are solo singing, group singing, and chanting songs, and Qupai (the names of the tunes to which Qu are composed) played by wind instruments, spiccato instruments and percussion instruments. Shangqing Ancient Music inherits and develops the essence of Qinghui school and Puan school Taoism music as well as Ci, traditional folk songs and folk music that have been popular in Taining since the Song Dynasty. Among them, most of the songs and wind music are unique to Taining and the surrounding areas, and also of great value for the study of local folk culture in Taining.

Taoist music, also known as ritual music, is a part of traditional Chinese music. It started around the time of the Northern and Southern Dynasties (420—589) and inherited the genes of ancient Chinese court music and traditional folk music. The Tang Dynasty was one of the heydays of the development of Taoism music. Emperor Xuanzong of the Tang Dynasty once made Taoist priests and ministers to offer Taoist music, and personally researched and taught Taoist music, which was used for memorial ceremonies for Lao-tzu in the Taiqing Palace. Emperors Taizong, Zhenzong and Huizong of the Song Dynasty respectively compiled dozens of Taoist music songs. Emperor Huizong of the Song Dynasty also held a national selection process to elect Taoist priests to learn Taoist music in Beijing. Emperor Zhu Yuanzhang of the Ming Dynasty established the Xuan institute to administer Taoism.

Taoism has a long history of development in Taining. As early as the end of the Western Han Dynasty, Wei Meifu from Nanchang, Jiangxi made alchemy by Shangqing Stream. Some local villagers worshiped him as a teacher and this custom passed on from generation to generation. Taoist music has also been passed down. However, the Shangqing Taoist Music is endangered because the elder generation of artists have passed away one after another. According to Li Jiqiu's recollection, among the materials left by his ancestors, there are more than ten original works about Taoism music. Among these books, there are more than 2,000 pieces of music and Qupai recorded in Gongchi Score (a traditional Chinese musical notation), however, there are only about 200 pieces of them left at present.

◎ 南禅宫。
The Nanchan Palace.

位于泰宁城南的南禅宫，是一个历史悠久、规模宏大的道观。据清乾隆版本的《泰宁县志》记载，南禅宫最早是南宋状元、宰辅邹应龙建的南谷书院中的南禅寺，是邹应龙子孙礼佛、祭祖的家庙。2008年，泰宁道教协会在这里成立，南禅寺就改为南禅宫。此后，这里每年都会有20场左右的法会集会，周边地区的摩郎仙云集在此。只要碰上这样的日子，就还可以在这里听到独特的道乐丝竹之声绵绵不绝。

The Nanchan Palace, located in the south of Taining Town, is a grand-scale Taoist temple with a long history. According to Taining County Records compiled in the reign of Emperor Qianlong of the Qing Dynasty, Nanchan Palace was originally the Nanchan Temple in Nangu (Southern Valley) Academy, which was set up by Zou Yinglong, who came first in the Southern Song Dynasty's highest imperial examination and became the former prime minister. It was the ancestral temple of Zou Yinglong's descendants and was used for worshiping Buddha and ancestors. In 2008, the Taining Taoist Association was established here, and Nanchan Temple was changed into Nanchan Palace. Since then, there are about 20 ritual gatherings every year, and Molang living in the surrounding area all gather here. As long as you encounter a day like this, you can also hear the unique sound from bamboo instruments of the Taoist music.

意外之美——忽如一夜大金湖

Beauty Caught by Surprise: Dajin Lake

◎ 美丽如画的大金湖。
The picturesque Dajin Lake.

◎ 大峡谷（翠谷雪花涧）。
The Grand Canyon.

20世纪初，美国著名作家杰克•伦敦写了一篇以阿拉斯加“淘金热”为背景的经典小说《赤金峡》。小说对淘金人在世外桃源般的峡谷中淘金的过程作了无比优美生动的描写，让全世界的读者对淘金人这个行当产生了浪漫的想象。

藏着金砂的峡谷是这样的：

“峡谷洁净无尘，绿叶与鲜花清新怡目，草地则像一块新织的天鹅绒……偶尔，一只蝴蝶在阳光与树影之间出没，到处响着野蜂的嗡嗡声，这些奢华的享乐者在花宴上熙熙攘攘，没有空闲相互争斗。小溪细细地流着，在峡谷里溜来绕去，沿途只是偶尔发出潺潺声。这声音仿佛梦中的呓语，时而因汇成水洼而寂寂无语，时而又因溪道狭挤而醒来叹息……在峡谷的这颗绿心里，一切都飘飘然。阳光与蝴蝶在树丛中飘来飘去。野蜂的嗡嗡与溪流的低语若隐若现。隐约之音与飘忽之色仿佛织成了一片缥缈的轻纱梦，而轻纱梦就是这里的地灵。它恬静却并非死寂，因而韵味悠长；它安宁而非无语，鲜活又不乱动；它憩息着，充满生机却没有苦难的搏斗。因此，这里的地灵是优雅的魂灵，是安适而又富裕的魂灵。”

At the beginning of the 20th century, Jack London, a famous American writer, wrote a classic novel *Burning Daylight*, setting in the "gold rush" in Alaska. The novel vividly described the process of digging gold in a paradise-like valley, which gave readers around the world a romantic imagination of the gold-digging industry.

The canyon with hidden gold sand is described as, "The canyon is clean without a dust, the green leaves and fresh flowers are eye soothing, and the grass is like a newly woven velvet … occasionally, a butterfly drifts through the rifted sunshine and tree shadow, there were wild bees all over flying around with buzz like the luxury hedonists, enjoy the banquet of blossom flowers with no time for fighting one another. The stream meanders through the canyon with occasional murmur. The sound is like the whisper in a dream, where it becomes silent because of the merger of water sometimes. Other times, it sighs as it awakens from the narrow squeeze of the channel … everything is so free in the green heart of the canyon. Butterflies drift through the rifted sunshine, and the buzz sound of wild bees and the murmur of streams come and go, the vague colours and sound form a dream like scene that becomes the soul of the place. It is soft but not silent, thus has a long-lasting charm; it is peaceful but not wordless, lively but not messy; it exists in tranquility, full of vitality without any conflicts. Therefore, the soul of the place is an elegant, peaceful and wealthy one."

◎ 溪流出峡谷。
The stream running through the canyon.

淘金人淘金的场景是这样的：

“他拿起了镐、铲子和淘金盆，敏捷地跳过水中一块块的石头，越过了池塘下游的小溪。在草坡、溪流相交处，挖出满满一铲的泥土，放进盆里。他蹲下来，两手拿着盆子，将它一半浸入水中，随即，灵活地转动着盆子，这样就可以使流水不断地淘洗盆中的泥沙……微小的金砂闪出来了。”

差不多是同一时期，在民国时期，很多江西人也翻越过武夷山脉，到泰宁城关的北溪一带，也就是金溪上游的支流溪谷中淘金。据说，这些省外淘金人的身影一直延续到了20世纪80年代。

没有太多留存的文字资料可以还原当时泰宁境内溪谷中淘金人的淘金场景。但在美国人杰克•伦敦的文字里，对世外桃源一般的幽静峡谷的描绘，却与泰宁的山水有着神奇的契合度。

杰克•伦敦笔下那洁净无尘的峡谷，峡谷里恬静缥缈的绿心，绿心里梦中吟语的溪水，仿佛也是在描绘泰宁。

泰宁也是一个自古产金子的地方。

泰宁古城在唐代时称“金城场”，就是因为此地产金；宋朝的《宋史・地理志》《闽产录异》《八闽通志》等也均有记载泰宁产金，是当时福建唯一向朝廷进贡黄金的县份；20世纪90年代，人们更是在泰宁城郊发现了何宝山金矿。

如果我们得以在空中俯瞰泰宁大地，就会惊叹于这片土地溪流密布、河沟纵横的样子。

泰宁是一个水养着的地方，不仅年降雨量大，而且小小境内遍布大小溪水河流73条。它们都汇入了金溪，而金溪之所以得名，就是因为溪中沙土含金。

因此，截流金溪之水形成的大金湖，也可以视为名副其实的黄金之湖。

◎ 淘金岁月。

The scene of gold-digging.

The scene of gold-diggers while digging gold is like this:"He picked up the pickaxe, the shovel, and the gold-digging pan, quickly leaped over the stones in the water and crossed the creek downstream of the pond. At the intersection of the grass slope and the stream, he dug a shovel of mud and put it in the basin. He knelt, holding the basin with both hands, immersing it in the water halfway, and then turning the bowl flexibly, so that the running water could wash the sediment in the basin continuously ... The tiny specks of gold flashed out."

Around the same time, many people from Jiangxi Province crossed Mount Wuyi to Beixi area during the Republic of China period, which is a branch of the upstream of Jinxi Stream, to dig gold in the valleys located near Taining Town. It is said that these gold-diggers were around until the 1980s.

◎ 淘金人。
A gold-digger.

There are not enough documents left to restore the gold-digging scene of the gold diggers in the valleys of Taining. However, the paradise-like canyons described by Jack London is a magical fit with the landscape of Taining.

The pure clean valley, the quiet and ethereal green heart of the valley, and the stream whispering in the dream of the green heart, depicted by Jack London, as if depicting Taining.

Taining is also a place for gold mining from ancient times.

Taining ancient city was called "Golden City Field" in the Tang Dynasty because of its gold production. *The Song (Dynasty) History: Geography* and some other books of the Song Dynasty also recorded the production of gold in Taining. Taining was the only county in Fujian Province that paid tribute to the court with gold at that time. In the 1990s, Hebao Mountain Gold Mine was discovered in the suburb of Taining.

If we overlook the land of Taining from the air, we would be amazed by the distribution of many streams and rivers in the area. It's a place rich in water. Not only it has a large amount of rainfall annually, but also has 73 rivers and streams throughout the small territory. They all merge into Jinxi (Golden Stream), which gets its name because the sand of the stream contains gold.

Therefore, Dajin Lake formed by the interception of the water of Jinxi Stream can also be regarded as a veritable golden lake.

大金湖的诞生

在1949年前，泰宁县没有通汽车，更没有铁路。

设想下，泰宁人要怎么去福建的省城福州？

如果坐船，一般在流经县城的杉溪上船。杉溪是泰宁的干流，在梅口乡与睢溪汇合，形成金溪，金溪过将乐，到顺昌，在顺昌城关又汇入闽江三大源流之一的富屯溪，然后到南平，又和其他两条闽江源流沙溪、建溪合流，进入闽江。泰宁人可在此换乘大船，一路顺溜而下抵达福州。

但在过去，这可是一段提着命在走的惊险万分的旅途。公元905年，晚唐诗人、陕西人韩偓受王审知邀请，从江西入闽，留下一首诗：

“长贪山水羡渔樵，自笑扬鞭趁早朝。

今日建溪惊恐后，李将军画也须烧。”

从这首诗的名字更容易理解作者的意思，诗名是《建溪滩波，心目惊眩，余平生溺奇境，今则畏怯不暇，因书二十八字》。这里的“建溪”，从他当时的行进路线看，应该指的就是闽北这一带崇山峻岭中的溪流。它们咆哮奔腾于深山峡谷间，滩多流急，连自觉胆大的陕西人韩偓都“心目惊眩”。

还有一首泰宁当地的民谣《芦庵滩头哭五更》：“四更鸡，声声泣，沉船浦里篙桨满，芦庵滩下万骨集。观音菩萨难救助，三官大帝叹无力。哭声三百里，处处招魂急……”芦庵滩就在泰宁金溪上游，是后来拦水筑坝的地方。

据说清光绪年间，泰宁金矿曾租给法国人开采，但20年租约未满，法国人就不来了，也是因为交通实在太费劲。

就是因为这样险恶的交通状况，泰宁一直深藏在闽北的深山之中，连带这令人惊艳的美景，也少为世人所知。

有文字可考，早在宋朝就有人开始赞美泰宁的山水了。宋朝宰相李纲，1130年到泰宁，称赞“山水之胜，冠于诸邑”。到了明朝，徐霞客几次游历福建，却都与泰宁失之交臂。但有一位叫池显方的文人，于1637年到过泰宁，留下了13首诗和2篇游记，算是泰宁山水最早的知音。到了清朝，江西僧人释最弱也为之吟咏：“怪石都从天上生，活如神鬼伴人行。海之内外佳山水，到此难容再作声。”

几千年来，仅有这几位文人骚客为泰宁留下一点声音，太少了。泰宁之美仿佛被上天刻意雪藏，跨过漫长的岁月，在等待一个属于它的时刻。

The Birth of Dajin Lake

Before 1949, Taining County did not have road access for cars, nor did it have railways. Imagine how people went from Taining to Fuzhou, the capital city of Fujian Province?

If people took a boat, they usually boarded a boat in Shanxi Stream, which was the main stream of Taining and flew through the county. It merged with Juxi Stream in Meikou Township, forming Jinxi Stream. Jinxi Stream flew through Jiangle, to Shunchang, and merged with Futun Stream, one of the three major branches of Min River at Shunchang town. Jinxi Stream then passed Nanping and merged with other two streams—Shaxi Stream and Jianxi Stream, and entered Min River. Taining people could transfer to big boats here and went down to Fuzhou all the way.

In the past, however, this was a dangerous and life-threatening journey. Some ancient poets described the route: The streams scream and run between valleys, they were speedy and had many twists and turns, even the self-recognized brave men like Han Wo from ShaanXi were "thrilled with his heart and eyes".

There is also a local folk song called *Cry on Lu'an Beach*, describing the danger of the place, "From one to three o'clock in the night, the crow of a rooster sounded like crying. The place where boats always sunk was full of paddles. Under Lu'an beach were the bones of the dead. The Goddess of Mercy could stop this, even the Jade Emperor and his officers could only sigh. The cries of the families

◎ 湖如明镜。
The mirror like lake.

of the dead could be heard from three hundred miles away, and they wanted to call the spirit of the dead back." Lu'an Beach was on the upper reaches of Jinxi Stream in Taining, where the dam was built to block water later.

It was said that during 1875 and 1908 in the Qing Dynasty, the French had rented the Taining gold mine. They had stopped coming to Taining before the 20-year lease ended because of the terrible traffic condition.

It was such treacherous traffic condition that made Taining hidden in the deep mountains of northern Fujian, the stunning scenery rarely known to the world.

It is recorded that as early as the Song Dynasty, some people praised the scenery of Taining. Li Gang, the primer minister of the Song Dynasty, arrived in Taining in 1130 and praised it "the mountains and rivers (of Taining) tops all the counties". In the Ming Dynasty, Xu Xiake visited Fujian several times but missed Taining. However, a scholar named Chi Xianfang came to Taining in 1637 and wrote 13 poems and 2 travelogues. He could be considered as the earliest bosom friend of the landscape of Taining.

Only these few literati and poets left their works about Taining throughout thousands of years, which was too little. The beauty of Taining seems to have been deliberately hidden by god and it has been waiting for the right moment that belongs to it as time passes by.

◎ 宏大的水上丹霞。
Marvelous Danxia landform over water.

◎ 池潭电厂大坝。
Chitan Dam.

1976年，闽江工程局在泰宁境内池潭村金溪流域的险滩——芦庵滩上拦河筑坝，截流蓄水。1980年大坝合龙，由此，险峻阻人到访的千沟万壑埋到了水下，换来了泰宁深山峡谷中一个长度达60余千米、湖面积5万多亩的人工湖，称为金溪新湖，简称金湖，现在称大金湖。

In 1976, Minjiang Engineering Bureau intercepted the river and built a dam on Lu'an Beach—a dangerous shoal in Jinxi Stream of Chitan Village in Taining. In 1980, when the dam closed, thousands of gullies were submerged in the water, forming a man-made lake with a total length of more than 60 kilometers and an area of 3300 ha in the deep mountain gorge of Taining, known as Jinxi New Lake, or Jinhu Lake for short. Now it is called Dajin Lake.

水漫过并密封了无数世代的奇峰异石，解除了泰宁山水的封印，又仿佛像武侠世界里一剂仙丹打通人的任督二脉一般，因为一湖碧水，人迹罕至的险峻崖谷成了绵延数十千米的水上丹霞景观——中国最大的水上丹霞奇观。

大金湖是目前福建最大的人工湖，也造就了福建面积最大的内陆湿地，犹如一颗明珠，镶嵌在闽西北的绵延山林之中。山水相映，美得曲折多致，美得浩瀚博大，美得幽深迷离，人们纷纷泛舟登临。

◎ 中国最大的水上丹霞奇观。
China's largest water Danxia spectacle.

Water has brimmed over the strange peaks and rocks of countless generations, removed the Taining landscape seal. The lake and the rare cliffs form a stretch of tens of kilometers of water Danxia landscape – China's largest water Danxia spectacle.

Dajin Lake is now the largest artificial lake in Fujian and it brings up the largest inland wetland in Fujian, which is like a pearl embedded in the continuous mountains in the northwestern Fujian. The waters and mountains reflect each other, making a variable and vast scenery, attracting many people to come for a cruise tour.

泛舟湖上，刘姥姥进了大观园

名人是非多，名湖水怪多。仿佛不是如此就不足以资世人津津乐道。中国的长白山天池、英国的尼斯湖等，世界上这些大名鼎鼎的湖，都有水怪的传说。大金湖也曾闹过“水怪”，还因此惊动了中央电视台来报道，当地甚至专门请了国内最厉害的捕捞队来抓“水怪”。后来，谜底揭开，原来是一种俗称“水老虎”的鳡鱼在作怪。

水怪不再是一个谜，但迷人的大金湖还是吸引了无数人从世界各地赶来领略它的风采。

游览大金湖要统一乘坐景区的游船，全程4到5个小时。外地游客若是坐动车来到泰宁，可在出动车站之后乘坐5路或者9路公交车到泰宁城区，然后在状元广场坐专线车，约25分钟就到大金湖了。

游船驶过平静的湖面，泛起层层涟漪，不时有白鹭在湖面翩翩起舞。

一座红红的丹霞山出现在你的眼前，主要景点有天迹水帘、野趣园、醴泉岩、尚书墓、甘露岩寺、鸳鸯湖、情侣峰、大赤壁、水上一线天、幽谷迷津、虎头寨、十里平湖、三剑峰、猫儿山等等。为了保护大金湖的自然生态，这些还只是大金湖风景中开放的一部分。百里湖区，水绕山前，山映水中，丹山碧水之妙，非亲历者不能与之语。

◎ 白鹭翩跹。
Flying egrets.

◎ 乘游船出发。
Cruising on the lake.

Cruising on the Lake

The gossips are to celebrities what the monsters in water are to famous lakes. Many famous lakes in the world like Tianchi at China's Changbai Mountain, Britain's Loch Ness all have legends of water monsters. Dajin Lake also had a "water monster", it even caught CCTV's attention to come to report it. The locals especially invited the best fishing team in the country to catch the "water monster". The mystery was unveiled later. The monster turned out to be a yellow-cheek carp. The water monster is no longer a mystery, but the fascinating Dajin Lake continues to attract people from all over the world.

To visit Dajin Lake, you need to take a cruise ship in the tourist area, which lasts 4 to 5 hours for the whole journey. Tourists taking bullet trains to Taining can take No. 5 or No. 9 bus to the city from outside the train station, and then take the special line at Zhuangyuan Square. From here, it will take about 25 minutes to arrive at Dajin Lake.

The ship sails across the calm lake and outlined pieces of ripples. From time to time, the egrets dance on the lake. Red Danxia mountains stand in front of your eyes. The main attractions are Tianji Shuilian (Sky Water Curtain), Yequ (Wildness-Fun) Garden, Liquan Rock, Shangshu Tomb, Ganlu Rock Temple, Yuanyang (Mandarin Duck) Lake, Qinglv (Lovers') Peak, the Grand Red Cliff, Yixiantian (A Sliver of Sky) Over Water, Valley Maze, Hutou (Tiger Head) Fortress, Shili Pinghu (the Calm Lake), Sanjian (Three-Sword) Peak, Maoer (Cat) Mountain and the like. In order to protect the natural ecology of Dajin Lake, these are only some of the open parts of the landscape. Those who have not experienced it just cannot imagine what a fascinating scene it is.

■ 大赤壁

大金湖最具标志性的是大赤壁，宽近500米，高约100米，兀立湖中，傲视四方。中国国家邮政局于2007年9月2日发行了《大金湖》特种邮票1套2枚，一枚的图案就是大赤壁，另一枚是猫儿山。

欣赏大赤壁，一般都是坐在游船上，贴壁而过，以仰视的角度，感叹它的伟岸、壮美。但如果不想像普通游客那样在游船上欣赏，还可以有另一种方式，就是登山对望，当然最好是选光影最魅的晨昏时候。

找一个早上，从县城出发，沿大金湖路开车，过了大金湖码头，差不多再开十几二十分钟，就到了泰宁县梅口乡江坑村附近一座叫马鞍山的山脚下。从这里开始登山，可以抵达大赤壁对面的山岗。碰上好的天气，可以看到大赤壁在湖水的环绕下，冷峻地耸立着，四周云雾缭绕，像是薄纱轻轻覆在美人身上，又像是神话中禅定的仙山。随着太阳慢慢升起，水面的雾缓缓升腾起来，聚拢到上方，延伸成环绕赤壁的一条长带，赤壁开始泛起朝阳的红光，变得艳红奇绝。长带善舞，变幻难测，赤壁也显露出各种身姿。最后，太阳完全出来了，云雾散去，赤壁露出它全部的真容，除了在湖上仰望之时的壮美之外，因为俯瞰这种角度的原因，也因为这样看到的赤壁是拥在水的怀抱之中，它多了一份柔媚恬静。

◎ 赤壁真容。
Getting close to the cliff.

◎ 大金湖大赤壁邮票。
The stamp of Dachibi.

Dachibi (The Grand Red Cliff)

The landmark of Dajin Lake is Dachibi, the largest red cliff, which is nearly 500 meters wide and about 100 meters high. It stands out from the rest of the lake. On September 2, 2007, China State Post Office issued a set of two special stamps for Dajin Lake.The pictures on the stamps are Dachibi and Maoer (Cat) Mountain.

Commonly, people view Dachibi by taking a cruise ship, sailing beside the cliff and look up so they can admire its magnificence. However, if you do not want to be like regular tourists to view the cliff from a ship, there is another way—to climb the mountains and enjoy the big red cliff from the opposite side. Of course, it is the best to choose the most charming period of the dawn or dusk with the fascinating condition of light and shades.

Departing from Taining Town in the morning, you can drive along Dajin Lake Road and pass Dajin Lake Dock to arrive at a mountain called Ma'an Mountain near Jiangkeng Village, Meikou Township in about ten to twenty minutes. From here, you can reach the mountain top opposite to Dachibi. If you are lucky enough to encounter good weather, you can see the Grand Red Cliff standing there surrounded by the lake and mist, like a beauty gently covered by a tulle or like a fairy mountain of mythology. As the sun rises slowly, the fog on the surface of the water rises as well, then gathers to the top and extends into a long belt around the red cliff. At that time, the red cliff becomes brighter and reflects the red light of the sun. With the changing and unpredictable fog, Dachibi shows various forms and postures. In the end, when the sun is completely out, the fogs are gone, Dachibi reveals its true appearance.

■ 甘露岩寺

泛舟大金湖，不可错过的还有甘露岩寺，这是与北岳恒山的悬空寺相媲美的。停船登岸，山势不高，路并不会太陡峭难走，拾阶而上，十来分钟便到了山门。山门呈拱形，由石头砌成，两侧有一副对联，上联为“层层楼阁一柱挈状元还母愿”，下联为“滴滴甘露万人饮观音赐吉祥”。未穿过山门时，只能看到偏殿的檐角，复行数十步，顿时豁然开朗。它建在下窄上宽、呈倒三角形的巨大岩穴中，中间那根粗大的红柱子支撑起整座寺庙，展现“一柱插地，不加（假）片瓦”的独具匠心的设计，整个寺庙像镶嵌在巨大的岩穴里一般。

◎ 岩洞里的甘露岩寺。
Ganlu Rock Temple built in the cave.

◎ 一柱擎寺。
The temple supported by one pillar.

Ganlu Rock Temple

One attraction you should not miss is Ganlu Rock Temple. It takes ten minutes to climb up the mountain along the stairs to arrive at the gate after getting off the cruise ship. The mountain is not high, so the road is not too steep. The stone gate is arched with a pair of couplets on both sides, and the scroll on the right reads "Layer upon layer of the building supported by a pillar, Zhuangyuan fulfilled his mother's wishes", and the one on the left reads "Drop by drop of sweet dew drunk by thousands of people, Guanyin (the Goddess of Mercy) bestows fortune". When you stand outside the gate, you can only see a corner of the temple, but after several steps going in, the view suddenly become extensive. The temple is built in a huge cave that looks like an inverted triangle. The thick red pillar in the middle supports the entire temple, so the temple seems to be embedded in a huge cave, showing the unique design of "an entire temple supported by one pillar that plunges into the ground with no added tiles".

■ 鸳鸯湖和情侣峰

鸳鸯湖，是大金湖的湖中湖，冬春季常有鸳鸯栖息于此，因此又是天下有情人的湖。

依偎水岸的，是一对恩爱的“情侣”，成就了永远的情侣峰。

■ Yuanyang Lake and Qinglv Peak

Yuanyang Lake, or the Mandarin Duck Lake, is inside Dajin Lake. Mandarin ducks are often found to inhabit here in spring and winter, hence the name and thus it is also known as a lake for lovers.

By the lake standing the Lovers' Peak, also called Qinglv Peak, looking like a couple of lovers deep in love are snuggling up to each other.

◎ 鸳鸯湖畔情侣峰。
Mandarin Duck Lake and Lovers' Peak.

水漈瀑布

水漈瀑布是大金湖最大的瀑布，发源于泰宁第二高峰——峨嵋峰的九栋岭，汇聚沿途的小溪、山泉注入大金湖，百米之外就能听到瀑布雷鸣般的响声。

令人称奇的是，在水漈瀑布左上角，远远看去就像有一尊观音水中打坐，手持净瓶，云水一天。

Shuiji Fall

Shuiji Fall is the largest waterfall in Dajin Lake, originating from the second highest peak in Taining—Emei Peak. It gathers streams and mountain springs along the way to merge into Dajin Lake. The thunder-like sound of the falls can be heard 100 meters away. And the image of Bodhisattva Guanyin on the upper left side makes the waterfall very special.

◎ 瀑布里的水帘观音。
An image of Bodhisattva Guanyin can be seen in the waterfall.

■ A Sliver of Sky over Water

Around the spot of A Sliver of Sky over Water, the canyon extends about 130 meters long, 100 meters high, four to five meters wide, and only 2.6 meters wide at its narrowest point. Passing through the valley on a ship, you only see "a slice of blue sky above and a slice of silver light (water) below".

◎ 水上一线天。
A Sliver of Sky over Water.

■ 水上一线天

水上一线天，谷长约130米，高近百米，宽四五米，最窄处仅2.6米。只觉“上有蓝天一线，下有银光一脉”。进入一线天，豁然一妙绝之笔。

◎ 水边观音。

An image of Bodhisattva Guanyin.

■ 幽谷迷津

幽谷迷津，旧称“二十四溪”，众多山溪曲折入溪谷，后来水库成湖，积水成潭，溪水半数以上已经成为湖中潜流，游船缓缓行驶在幽静的溪谷中，隐隐还可以看到当年的石阶山道。

■ The Valley Maze

The Valley Maze used to be called "Twenty-four Streams" because there were many mountain streams that ran into the stream valleys. Later, the reservoir was built and formed Dajin Lake, so more than half of the streams became undercurrent in the lake. When sailing slowly on a ship along the streams inside the peaceful valley, you can faintly see the stone steps of the mountain paths from ages ago.

◎ 幽谷迷津。
The Valley Maze.

■ 十里平湖

十里平湖，南北长约5千米，东西宽约2千米。在这里，不同天气下有不同的气象景观。晴天，风和日丽，波光粼粼；雨天，波涛汹涌，山色凄迷；还有山雾、日出、晚霞、星月等许多变幻莫测的迷人景象，正如苏轼诗句所言："水光潋滟晴方好，山色空蒙雨亦奇。"

■ The Calm Lake

The Calm Lake (Shili Pinghu) is 5 kilometres long from north to south and 2 kilometres wide from east to west. There are different views under different weather conditions here. On sunny days, a breeze ripples the surface of the lake while the sunshine makes it sparkling. When it rains, the waves run high and the colour of the mountains are blurred. Plus, the mountain mist, sunrise, sunset glow, the moon and the stars, and many other unpredictable charming scenes make the lake a fascinating place.

◎ 十里平湖。
The Calm Lake.

看它的万种风情

一般看大金湖，是在天气晴朗的白天，但大金湖有千面的壮美，万种的风情。要领略到这些，需要时运，也需要你有更敏锐的眼睛和更浪漫的心境。如范仲淹在《岳阳楼记》中描述的洞庭湖一般，有着万千气象等着你去邂逅。霪雨霏霏时，日星隐曜，山岳潜形，满目萧然；春和景明时，上下天光，一碧万顷，心旷神怡；皓月千里，又是浮光跃金，静影沉璧，渔歌互答，此乐何极！

An Ever-Changing Gallery of Spectacles

Generally speaking, it is the best to visit Dajin Lake during daytime on bright sunny days. However, Dajin Lake has many charming sides to offer. It presents an ever-changing gallery of spectacles waiting for you to explore. Some may come during a spell of rains, which may drag on from month to month, blocking the rays of the sun and stars, and hiding the mountains from sight. The forlorn sight will move people to grief. Yet others may come in warm spring and enjoy a crystal-clear view of the unruffled lake, which reflects the sky like a gigantic iridescent mirror. Or they may see the bright moon casting her beautiful light far and wide onto the shimmering lake, reflecting a silver glittering upon the water like a piece of jade lying at the bottom of water, and feel extremely happy when hearing fishermen's interactive songs under the moonlight.

◎ 第一缕阳光。

The first ray of sunshine.

◎ 雨后彩虹。
A rainbow over the lake after rain.

如果能在大金湖遇到一阵急雨，那是它格外给你的礼物。山雨欲来惊奇，雨中苍茫，雨后更媚。尤其在夏日，一场猝不及防的雨，会把大金湖洗得更加清澈，呼吸雨后的空气，也有了润到心肺的感觉，而一番喧腾后的大金湖，其宁静深沉更容易为人感知，江流有声，断岸千尺，山高人小，人仿若融化在一个神清气朗、大美无言的天地里。

If you encounter a torrential rain, it is a gift presented especially for you by the lake. The scenery is amazing before rain in the mountains. It is boundless during the rain, and enchanting after the rain. In summer particularly, sudden rainfalls make the water in Dajin Lake clearer. The air after rain can be quite refreshing for your heart and lungs. It is easier to feel the peace and tranquility of Dajin Lake after the noisy rain. With running river, thousand-feet shore, and high mountains surrounded, one may feel like melted in a heavenly and wonderful world.

还有大金湖的黄昏，是最醉人的。太阳慢慢向水漈瀑布山后滑落，原本碧蓝的天空旋即变成微红，然后是酡红。白云也变得辉煌，先是一朵一朵，而后以各种美妙莫测的形状慢慢地铺展开了，层层晕染。太阳完全落到山后去了，落日熔金，火光烧红了天际，湖水映照着这片火红金黄，水天一色。然后，太阳真的隐退了，沉到了世界的另外一边。暮归的飞鸟开始划过静静的天空，静默地书写生灵的律动，呈现出一种澄净的美感。

The most intoxicating time of Dajin Lake is its dusk. The sun slowly slides down towards Shuiji Fall, the blue sky turns slightly red and then dark red. White clouds also become brilliant, firstly like flowers, then slowly extend with changeable shapes, layer by layer. When the sun completely falls behind the mountain, the remaining light of the sunset is like melting gold, the brightness lightens up to the sky edges. The lake reflects the sunset and its fire-like red and golden yellow colors, the water and the sky have become one. Finally, the sun disappears and falls to the other side of the world. The flying birds begin to cross the quiet sky on their way home, setting the rhythm of the lively nature in a peaceful manner, bringing up a purified beauty.

◎ 渔歌唱晚。
Returning of the fishermen at sunset.

◎ 全国最美赏月地。

The most beautiful place for moon-watching in China.

夜观大金湖，山行起伏，恬淡安然，让人忍不住又想起苏东坡的词："清风徐来，水波不兴。举酒属客，诵明月之诗，歌窈窕之章。少焉，月出于东山之上，徘徊于斗牛之间。白露横江，水光接天。纵一苇之所如，凌万顷之茫然。"2013年，中央电视台新闻中心官方微博曾发起投票，福建泰宁大金湖被票选为"全国最美赏月地"。当年的中秋之夜，央视新闻频道直播了泰宁"中秋最美赏月地"——大金湖畔的赏月现场。

The quiet night view of Dajin Lake, with mountains up and down, reminds people of the poem of Su Dongpo, "There was a fresh, gentle breeze, but the water was unruffled. I raised my winecup to drink to my friends, and we chanted the poem on the bright moon, singing the stanza about the fair maid. Soon the moon rose above the East Mountain, hovering between the Dipper and the Cowherd. The river stretched white, sparkling as if with dew, its glimmering water merging with the sky. We let our craft drift over the boundless expanse of water, feeling as free as if we were riding the wind bound for some unknown destination, as light as if we had left the human world and become winged immortals (By Yang Xianyi, Gladys Yang)." In 2013, Dajin Lake was voted as the most beautiful place for moon-watching in China by the official Weibo of CCTV News Center. On the Mid-Autumn Day of the year, the moon-watching site at Dajin Lake was broadcast live on CCTV News Channel.

◎ 金秋，层林尽染，如诗如画。

The picturesque autumn when the forests, mountains and water make a spectacle like oil paintings.

◎ 冬日，“红装”素裹，分外妖娆。

The red cliffs look more beautiful than ever when covered with white snow in winter.

有山临湖相照

水依山，山傍水。可在水上看山，也可从山上看水。

大金湖边上，最有名的景点之一是猫儿山。要从大金湖上看猫儿山，得跨越烟波浩渺的十里平湖。猫儿山陡立于湖中，迎面而至，仿佛在群山中窥视着湖上看它的人。

猫儿山是远眺大金湖的最佳地点。要去猫儿山上看大金湖，得坐车，会先经过大金湖悬索桥。这时远远就能望见湖中三座山峰直插苍穹，这是猫儿山又叫三剑锋的缘故。三剑峰是丹霞地貌中最具特色的峰丛景观，岩层在地球引力作用下产生垂直裂隙，把岩层切割成一个个柱体，地壳抬升后，经流水侵蚀，周围的柱体崩塌，只有三个高达150余米的岩柱保留下来。

继续沿着绕山公路前行，角度随之变化，再回望三剑锋，它已变幻成一只猫的样子蹲坐着，望着大金湖的十里平湖。

◎ 2007年发行的猫儿山邮票。
The stamp of the Cat Mountain issued in 2007.

◎ 金猫望湖。
The giant cat looking at the lake.

◎ 三剑峰。
The Three-Sword Peak.

Mountains by Dajin Lake

The continuous mountains and surrounded water allow us to see the mountains from the water and view the water from the mountains.

Maoer (Cat) Mountain is one of the most famous attractions around Dajin Lake. To watch Cat Mountain from the lake, you need to come to the vast Calm Lake. You will then see Cat Mountain standing steeply in the lake and peering at visitors come toward it.

At the same time, Cat Mountain is the best place for overlooking Dajin Lake. You must take bus there. When the bus passes Dajin Lake suspension bridge, you can see three peaks like swords pointing straightly into the sky. This is why Cat Mountain is also called the Three-Sword Peak. The Three-Sword Peak is the most characteristic peak-cluster landscape of Danxia landform. The rock layers produced vertical fissure under the gravitational force of the earth, and the rock layers were cut into columns. After the earth crust had been lifted up, the surrounding columns collapsed because of the erosion of running water. Only three 150-meter high rock pillars have remained.

The viewing angles change along with the winding road. If you look back to the Three-Sword Peak at this time, the view of the peak has changed to the image of a giant crouching-cat that is looking at the Calm Lake.

◎ 猫儿山顶观日出。
Watching the sun rising at the top of Cat Mountain.

去往猫儿山山顶，要从山后一条石阶山道步行而上。经曲径入山，可见苍藤老树，林荫筛风，山花藐藐，幽静动人，让你沉默不语。待抵达山顶，带着微喘的呼吸和微微出汗的身体，你突然就沦陷在一片开阔无边的山光水色之中，眼前是白云，白云之下星罗棋布的小岛散落在碧蓝的水中。极目远眺，这时呼吸畅快了，感觉在和天地同呼吸，身体也清爽了。一丝丝不知道从哪里冒出的清爽的风钻进身体，感觉整个人像要飞起来一般，正如苏东坡一样，“划然长啸，草木震动，山鸣谷应，风起水涌”。

To get to the top of Cat Mountain, it is necessary to climb a set of stone stairs along a trail behind the mountain. Along the winding trail, you can see various old trees with dense branches and smell fragrance of different kinds of mountain flowers. The tranquility and quietness will make you stay in silence to feel the peace and calmness. When you arrive at the mountain top, with a slight pant and slight sweat, you are suddenly being immersed in a vast expense of mountains and waters. The white clouds float in the sky, and small islands scatter in the water. Looking into the distance, your breath suddenly becomes smoother, and you'll feel as if you are breathing at the same rhythm as the heaven and the earth. The cool and refreshing wind gets through your clothes and makes you feel as if you are about to take off and fly.

和三剑峰隔湖相望的是黄石寨。它的入口就在大金湖悬索桥畔一个叫“小赤壁”的断崖绝壁上，它所在的山也叫黄石山。在黄石山上可以看到北面的十里平湖和南面的戈口平湖，黄石寨就扼守在大金湖这两块最大湖区的咽喉处。黄石寨右下侧陡壁上有个仅能容纳10多人的小岩洞，叫“人干岩”。这就是明朝泰宁籍进士、官至四品太仆寺少卿的江日彩少年时候的读书处，后来改叫“光台”，在景区宣传册上，这里叫作“光台日照”景点。关于江日彩，最为人知的应该是他推荐邵武知府袁崇焕出守辽东御敌的故事；少为人知的是，清军入关后，明朝覆灭，他的儿子江豫、江夏率几万民众避乱世于泰宁的石辋南石寨，设隘自卫，最后石寨陷落，万人惨遭杀戮。

On the other side of the lake, opposite to the Three-Sword Peak is Huangshi (Yellow Stones) Fortress. Its entrance is on a cliff called Xiaochibi (Small Red Cliff) beside the suspension bridge, thus the mountain where Huangshi Fortress is located is also called Huangshi Mountain. From the mountain you can see the Calm Lake on the north and the Gekou Pinghu Lake on the south. Huangshi Fortress holds a strategic point of the two largest lake areas of Dajin Lake. There is a small cave in the steep wall at the lower right side of Huangshi Fortress, which can only fit around 10 people at a time. This was the study site of young Jiang Ricai from Taining, who was awarded "*Jinshi*" (one of the top rankings) of an imperial examination in the Ming Dynasty.

◎ 大金湖悬索桥畔“小赤壁”。
The suspension bridge and the Small Red Cliff.

◎ 从十里平湖远眺虎头寨。
Looking at Hutou (Tiger Head) Fortress from the Calm Lake.

在十里平湖的西侧，还有一座始建于宋代的军事要塞——虎头寨。寨子“怪石巉岩如虎头，然山巅平坦，流泉松竹，时振海涛”，据说原有房舍田产、池塘水井，可供上千人住几个月。虎头寨在杉溪和濉溪的交汇处，扼守通往闽江的重要门户。元朝末年，与朱元璋争夺天下的陈友谅曾率兵在此驻扎。此外，传说太平天国翼王石达开的部下由于率军南下受阻，也曾在此安营扎寨。之后，虎头寨慢慢变成了文人墨客垂青流连的胜地。清朝初年，这里曾有多处石建筑，但现在都已毁坏，只剩棋盘石等若干遗址。

虎头寨正好在十里平湖最旷阔的中段，寨前有一巨石，似巨虎仰天长啸，人称虎头啸石。站在高出湖面100米的虎头寨，你就拥有一览无余的视野，万顷烟波，尽收眼底。

On the west side of the Calm Lake sits an important military fortress of the Song Dynasty—Hutou (Tiger Head) Fortress. The landscape of the fortress is made up of strange rocks and stones and the shape of it is like the head of a tiger. However, there were springs, bamboos and pine trees inside the fortress. It was said that the houses, farmlands, ponds and wells here in ancient times were plenty enough for thousands of people to live for a few months. In the early years of the Qing Dynasty, there were many stone buildings, but all had been destroyed, leaving only Qipanshi (Chessboard Stone) and a few other historical sites.

Tiger Head Fortress is located right in the middle of the widest area of the Calm Lake. There stands a huge stone in front of the fortress like a giant tiger screaming to the sky. Therefore, people named it the Tiger Roar Stone. If you stand at the Tiger Head Fortress that is 100 meters above the lake, you can enjoy a panoramic view of the vast expanse of mountains and lakes.

◎ 从虎头寨俯瞰大金湖。
A bird's eye view of Dajin Lake from Tiger Head Fortress.

迷路峡谷——领略中国丹霞青年期的样子

2010年8月2日，在巴西利亚举行的第34届世界遗产大会上，经联合国教科文组织世界遗产委员会批准，“中国丹霞”被正式列入《世界遗产名录》。

“中国丹霞”是一个系列提名的世界自然遗产项目，由贵州赤水、福建泰宁、湖南崀山、广东丹霞山、江西龙虎山（龟峰）、浙江江郎山6个中国亚热带湿润区著名的丹霞地貌景区组成。

山和人一样，都会经历出生、成长、衰亡，丹霞地貌依据其侵蚀与发育过程，分为青年期、壮年期、老年期。泰宁丹霞，正是一个丹霞地貌单元发育周期中处于风华正茂的时期。

贵州赤水：青年早期——强抬升、深切割高原峡谷型丹霞的代表；

福建泰宁：青年晚期——深切割、山原峡谷曲流和多成因崖壁洞穴的代表；

湖南崀山：壮年早期——密集型圆顶、锥状丹霞峰丛峰林的代表；

广东丹霞山：壮年晚期——簇群式丹霞峰丛峰林的代表；

江西龙虎山（龟峰）：老年早期——疏散型丹霞峰林与孤峰群的代表；

浙江江郎山：老年晚期——高位孤峰型丹霞地貌的代表。

◎ 泰宁密集的网状峡谷曲流。

Reticulate deep mountain plateau canyons and streams in Taining.

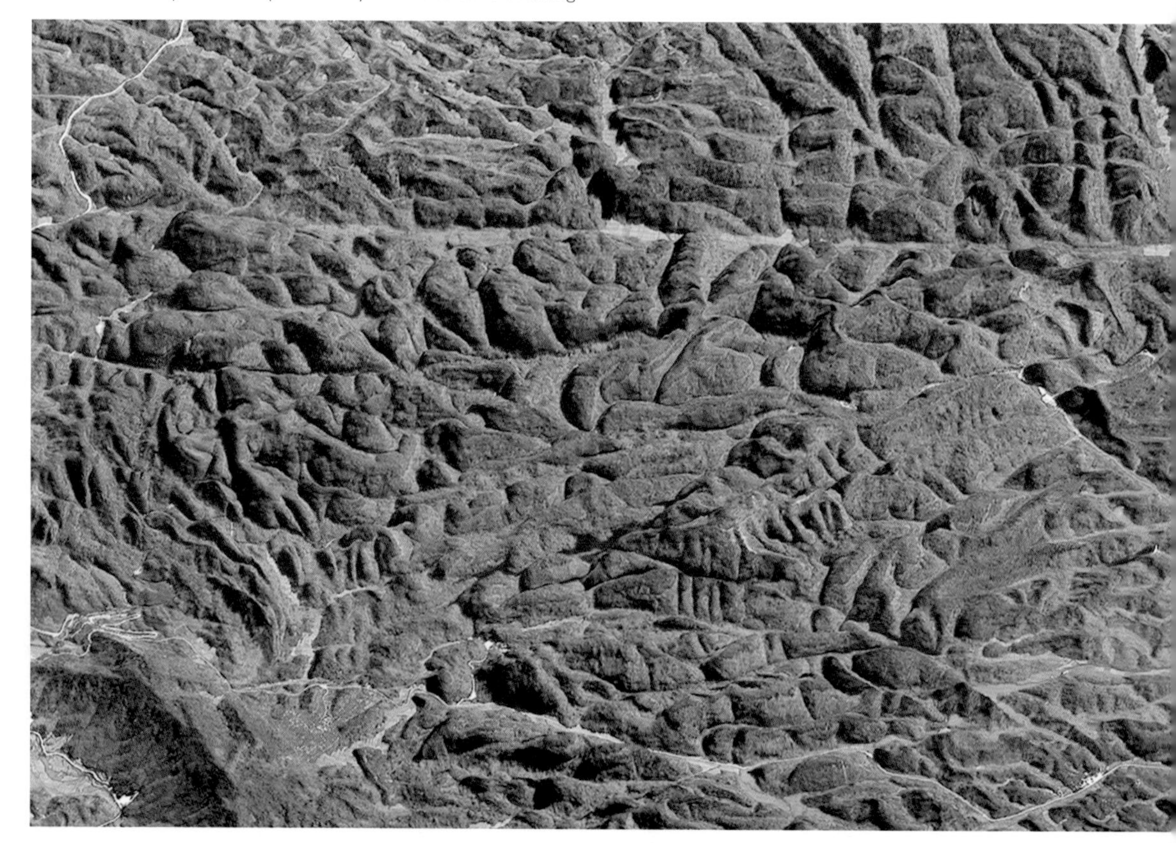

Getting lost in Canyon: Experiencing the Adolescence of Chinese Danxia

At the 34th World Heritage Convention held in Brasilia on August 2, 2010, "China Danxia" was officially listed on the World Heritage List.

China Danxia consists of 6 famous Danxia landform tourist areas in the humid subtropical zone of China, namely, Chishui in Guizhou Province, Taining in Fujian Province, Mount Langshan in Hunan Province, Mount Danxia in Guangdong Province, Turtle Peak on Mount Longhu in Jiangxi Province, and Mount Jianglang in Zhejiang Province. Mountains, like human beings, will experience birth, growth, maturity, ageing and death. According to its development processes, Danxia landforms go through infancy, youth, prime period and ageing period. Once a new round of erosion occurs, the mountain rises again and Danxia regenerates, and there will be a new round of growth. Taining Danxia is now in the peak time of its youth period.

Chishui in Guizhou Province: Early stage of the youth period—the representative of high elevation, topographic uplifting deep cutting Danxia canyons.

Taining in Fujian Province: Late stage of the youth period—the representative of deep mountain plateau canyons and multi-genesis cliff caves.

Mount Langshan in Hunan Province: Early stage of the prime period—the representative of peak clusters with intensive-dome or cone-shaped peaks.

Mount Danxia in Guangdong Province: Late stage of the prime period—the representative of cluster-type Danxia peaks.

Mount Longhu (Turtle Peak) in Jiangxi Province: Early stage of the senior period—the representative of spread-out Danxia peak forests and solitary peak groups.

Mount Jianglang in Zhejiang Province: Late stage of the senior period—the representative of high solitary peaks of Danxia landform.

◎ 完好的古夷平面。
Intact ancient planation surface.

对泰宁丹霞遗产地的价值是这样描述的："泰宁丹霞地貌区保存了清晰的古剥夷面，被密集的网状峡谷和巷谷分割为破碎的山原面；独特的崖壁洞穴群、密集的深切峡谷曲流和原始的沟谷生态构成其罕见的地质遗迹特征，成为青年期低海拔山原峡谷型丹霞地貌的代表；峡谷急流与密集峰丛紧密结合，山水景观优美；保持了生态环境的原生性、生物和生态多样性。"

有调查表明，泰宁丹霞地貌区有深切峡谷470多条，其中巷谷150余处，线谷80余处，无论峡谷数量、密度、窄度和深切峡谷曲流的曲度以及组合形态之复杂多变，都是独一无二的；大型单体丹霞洞穴60多处，壁龛状洞穴群100多处，千姿百态的大小洞穴则难以计数。

简而言之，泰宁丹霞以"密集的网状谷地、发育的崖壁洞穴、完好的古夷平面、宏大的水上丹霞"，在"中国丹霞"系列遗产地中具有不可替代的地位，被冠以"水上丹霞园""峡谷大观园""洞穴博物馆"等种种奇观美誉。

The nomination of Taining Danxia is described as follows, "The Taining Danxia landform area maintains a clear ancient planation surface and an original mountain face which is divided into fragmentized surface by dense netlike canyons and U-shaped valleys. The unique cliff wall cave groups, the dense deep cutting canyons and meandering streams, and primitive gully ecology have formed rare natural features. It is the representative of low-altitude mountains and gorges of adolescent Danxia. The rapid streams in the canyon intertwine with dense peaks, which shapes the beautiful scenery of the mountains and streams. The primordial ecological environment and the bio-diversity are well preserved here."

According to relevant research, there are more than 470 deep cutting canyons in Taining, which includes more than 150 U-shaped canyons and over 80 V-shaped canyons. Taining Danxia is unique not only because of the total number of canyons, their density, their narrow width and the curvature of meanders, but also because of their complex and varied shapes and forms. There are over 60 large single Danxia caves, more than 100 wall-like cave groups and countless caves that differ in their size and shapes.

Simply put, Taining Danxia boasts "the most dense net-like valleys, the most full-grown cliff caves, the intact ancient planation surface, and the most magnificent Danxia landform on water". Taining Danxia plays an irreplaceable role in "China Danxia". It has many spectacle labels, such as "Water Danxia Garden", "Grand View Canyon Park" and "Cave Museum".

丹霞大发现——从地质博物馆看起

“丹霞”一词源自曹丕的《芙蓉池作诗》中的“丹霞夹明月，华星出云间”，指天上的彩霞。这是一个绚烂美艳的词语，但作为一种地貌类型，长久以来在很多人眼里却并不可爱，在泰宁当地人看来，这不过是令人讨厌的“红石山”，是穷山恶水，种不出粮食，也不能就地取材用来盖房子，简直无一用处。

“丹霞地貌”也叫“红层地貌”，这个名词诞生至今，还不到100年的时间。

1928年，中国地理学家冯景兰，在广东丹霞山注意到了分布广泛的第三纪（6500万—165万年前）红色砂砾岩层，在长期风化、侵蚀和重力作用下，形成了堡垒状的山峰和峰丛等千姿百态的地形，把它命名为“丹霞层”。

1938年中国地质学家陈国达首次提出“丹霞山地形”的概念，1939年他正式使用了“丹霞地形”这一分类学名词，以后丹霞层、丹霞地形（地形即地貌）的概念便被沿用下来。

经过几代地理学家的研究，中国越来越多的丹霞地貌被发现，目前已查明丹霞地貌1005处，在我国热带、亚热带湿润区，温带湿润—半湿润、半干旱—干旱区和青藏高原高寒区的28个省均有分布。

◎ 丹霞出云间。

A giant Danxia mesa under the clouds.

The Discovery of Danxia — Beginning the Journey from the Geological Museum

The word "Danxia" originates from a poem written by Cao Pi. In *Writing a Poem by the Lotus Pond*, he wrote, "A bright moon rises amid the Danxia, stars flash through the clouds", where Danxia refers to roseate clouds in the sky. Danxia is a glamorous word, but as a landform, it has long been disliked by many. To Taining locals, Danxia is just annoying "Red Rock Mountains"—they are poor mountains and bad torrential rivers because they cannot be used as farms to grow wheats nor can they be used for building houses; they seem useless.

The term "Danxia landform", also known as "red beds", has been used for less than 100 years.

Feng Jinglan, a well-known Chinese geographer, noticed a wide distribution of red glutenite layers in the Tertiary (65 million to 1.65 million years ago) at Mount Danxia in Guangdong Province. Long-term weathering, erosion, and the act of gravity resulted in bastion-like peaks and peak forests. Feng Jinglan realized that this was a unique landscape and named it the "Danxia Layer" in 1928.

In 1938, Chinese geologist Chen Guoda first put forward the concept of the Danxia Mountain Landform. In 1939, he formally used the taxonomic term Danxia Landform, and later the concepts of the Danxia Layer and Danxia Landform were adopted ever since.

More and more Danxia landforms have been discovered in China attributed to the research that has been done by several generations of geographers. There are 1,005 sites of Danxia landforms that have been identified until now, which are distributed across 28 provinces of China's tropical and subtropical humid regions, temperate humid—sub-humid, semi-arid—arid regions and the Alpine regions of the Qinghai-Tibet Plateau.

◎ 绵延雄浑的早期峰丛。
Large scale of cone-shaped peaks.

中国各地的丹霞地貌风景差异很大：西北丹霞辽阔壮丽，有青海坎布拉、宁夏须弥山等；西南丹霞地形起伏剧烈，高耸的赤壁丹崖常与急流瀑布相伴而生，气势磅礴，有贵州赤水、四川青城山、云南老君山等；东南丹霞，主要就是指泰宁、江郎山、龙虎山、武夷山、冠豸山、丹霞山、桃源洞、齐云山、方岩、万佛山等著名丹霞风景，以深峡、奇峰、断崖、岩穴为特点。

泰宁丹霞地貌面积约252平方千米，列入遗产地核心区域的范围为110平方千米，居中国东南各省丹霞地貌之冠。

The Danxia landforms vary greatly from place to place: the northwest Danxia is vast and magnificent, examples include Kanbula of Qinghai and Mount Xumi in Ningxia; the southwest Danxia terrain is undulating, where the towering Danxia cliffs have often been formed together with rapid waterfalls with amazing grand views, such as Chishui in Guizhou, Mount Qingcheng in Sichuan, Mount Laojun in Yunnan. Danxia in the southeast mainly refers to the famous Danxia scenery like Taining, Mount Jianglang, Mount Longhu, Mount Wuyi, Mount Guanzhai, Mount Danxia and Taoyuan Cave..., characterized by deep valleys, wonderful peaks, bluffs and caves.

The area of Danxia landform in Taining is about 252 square kilometers, ranking the top of Danxia landforms in the southeastern provinces of China.

◎ 泰宁丹霞峰柱。
A Danxia pillar in Taining.

◎ 大赤壁。
Dachibi (the Grand Red Cliff).

丹霞地貌分为正地貌、负地貌和微地貌。由于漫长地质时期流水侵蚀、风化剥蚀、重力崩塌等外力作用缓慢雕塑，发育出丹霞崖壁、方山、石峰、石柱、石墙、崩堆积等正地貌。比如泰宁的丹霞正地貌可以见到以下几种：

■ 丹霞崖壁

坡度大于60度、高度大于10米的陡崖坡，平地拔起，巍然高耸，比如金湖上的大赤壁。

In geomorphology, there are three major categories of Danxia landforms: positive landform, negative landform and micro-landform. The positive landform is a topographic form resulting from erosion, weathering, the action of gravity. Such external forces gradually carve, projecting notably upward to form cliffs, mesas, peaks, pillars, walls or colluvia. The following are some examples of positive Danxia landforms in Taining:

■ Danxia Cliffs

It is a kind of steep cliff with a slope that is greater than 60 degrees and a height that is over 10 meters. It rises straight from the ground and soars high. Dachibi on Dajin Lake is an example.

丹霞方山

丹霞方山山顶平缓，四壁陡立，呈城堡状，往往只有一条小道可通达山顶。古人常利用方山筑寨，如黄石寨。

Danxia Mesas

Danxia mesas have flat tops and each has four steep sides, which look like castles. Each mesa usually only has one trail as access to the top. In ancient times, people often built villages or fortresses in these square-shaped mesas, such as the famous Huangshi (Yellow Stone) Fortress.

◎ 黄石寨。
Huangshi Fortress.

■ 石峰石柱

泰宁丹霞地貌中有着发育良好的红层奇观——丹霞石峰、石柱。奇峰异柱，争奇斗艳。有的似情侣相依，如情侣峰；有的如仙人相聚，如八仙崖。

■ Danxia Peaks and Pillars

Taining Danxia contains spectacular red-beds landform that are well-developed, such as isolated peaks, towers and pillars. They are shaped in different forms and figures as if they try to compete for attention. The Lovers' Peak, for example, looks like two love birds snuggling up to each other. Baxianya (the Eight-Immortal Peaks) remind us a picture of several immortals gathering together.

◎ 八仙崖。
The Eight-Immortal Peaks.

◎ 月老岩。
Yuelao Rock (Laojun Rock).

■ 丹霞石墙

薄薄的呈长条状的山体，长度比宽度大两倍甚至数十倍，高度也大于宽度，像是一堵凌空降下的墙，壮观异常，如月老岩（也有人称之为老君岩）。

■ Danxia Stone Wall

This is a kind of thin and strip-shaped mountain with a length that is two to ten times larger than the width, and with a height that is greater than the width. It looks like a wall towering to the skies and the view is spectacular. Yuelao Rock (also called Laojun Rock) is a good example of the Danxia stone wall.

◎ 云崖岭崩积巨石。
Colluvial stones at Yunya Ridge.

■ 崩积堆和崩积巨石

多在陡崖之下，因崖壁重力崩塌堆积而成，或堆积成山丘，或形成崩积洞穴，如寨下大峡谷的云崖岭。

■ Colluvia and Colluvial Giant Stones

The colluvia are mostly piled up or caved at the base of a cliff or slope. The reason for the formation of colluvia is the effect of gravity. The giant stones on Yunya Ridge in the Zhaixia Grand Canyon is a kind of colluvial stones.

泰宁丹霞地貌中常见的有这么几种负地貌：

■ 丹霞沟谷

这是丹霞负地貌中最重要的景观，包括宽谷、峡谷、巷谷、线谷等，如一线天；因受流水侵蚀深切形成的曲流，如上清溪。

There are several negative landforms in Taining:

■ Danxia Valleys

This is the most important landscape of Danxia negative landforms, including wide valleys, canyons, ravines, U-shaped canyons, V-shaped canyons, such as a Sliver of Sky. Shangqing Stream is a meandered stream that has formed because of deep-cutting of water.

◎ 上清溪峡谷。
A canyon over Shangqing Stream.

◎ 岩洞里的甘露岩寺。
Ganlu Rock Temple in the cave.

丹霞洞穴

丹霞洞穴成洞的部位多在软弱的岩层，软岩层相对快速的风化使之凹进，形成洞穴，主要有额状洞、扁平洞、拱形洞等类型，如状元岩、丰岩、甘露岩上的岩洞。

Danxia Caves

Danxia caves are mostly formed in soft rock layers. The relatively rapid weathering makes the soft rock layer recess to form caves. These caves are mainly frontal caves, flat caves and other types such as the caves in Zhuangyuan Rock, Fengyan Rock and Ganlu Rock.

■ 丹霞穿洞

其实这也是丹霞洞穴的一种。在丹霞岩层中，软岩地段先形成洞穴，日久天长，石墙被蚀穿，两边通透，就成了穿岩。穿岩如果被侵蚀，岩壁后退并不断垮塌，当跨度大于洞顶厚度的时候，就成为石拱，石拱横跨在河谷上的称天生桥，如大田读书山的天生桥。

■ Danxia Arches

This is also a kind of Danxia caves. The soft rock sections of the thin Danxia walls first form caves and over time, the thin walls are penetrated under the effect of erosion. An arch is formed when the two sides of the walls are transparent. If the rock is eroded, and when the span is longer than the thickness of the top of the cave, it becomes a stone arch. The stone arch across the river valley is called Tiansheng (means from the sky) Bridge, such as the Tiansheng Bridge in Datian Township.

◎ 大田乡的天生桥。
The Tiansheng Bridge in Datian Township.

◎ 天穹岩。
The Sky-Dome Rock.

■ 丹霞微地貌

丹霞地貌中还有一种蜂窝状洞穴，是在崖壁上或洞穴顶部星罗棋布的细小洞穴，如同剖开的蜂房，如寨下大峡谷的天穹岩，属于丹霞微地貌的特殊景观。

■ Danxia Micro-landform

There is also a kind of beehive-shaped cave, which are clusters of small-caves on the cliff or on top of the cave, like an open hive. An example of the beehive-shaped cave is Tianqiong (the Sky-Dome) Rock of the Zhaixia Grand Canyon. This kind of cave is a special micro-landform of Danxia.

丹霞是泰宁最耀眼的地理景观标志，但不是唯一。早在“中国丹霞”成为世界自然遗产的前5年，即2005年，泰宁就被联合国教科文组织公布为第二批世界地质公园，由石辋、大金湖、八仙崖、金铙山四个园区和泰宁古城游览区组成。泰宁世界地质公园以典型青年期丹霞地貌为主体，兼有火山岩、花岗岩、构造地貌等多种地质遗迹，是一个综合性地质公园。

因此，在你目睹泰宁山水之前，建议你先去参观一个地方，那就是泰宁地质博物馆。它位于距泰宁城关2千米处，里面有泰宁的缩微模型，上面标有泰宁所有的景点，还配有详细的文字介绍。参观完地质博物馆，你会对泰宁世界地质公园的概况、形成背景及典型的丹霞景观有一个总体认识。对泰宁即将呈现给你的大自然之美有了初步了解之后，你就可以出发了。

最后，还想提醒你的是，欣赏丹霞需要一种诗意的心境。

丹霞，是一种专为浪漫的审美而生的地貌。

Danxia is the most dazzling geographical landmark of Taining, but not the only one. Taining was declared by UNESCO as a Global Geopark in 2005—five years before "China Danxia" became the world's natural heritage site. It consists of four parks including Shiwang, Dajin Lake, Baxianya (Eight-Immortal Peaks), Jinnao Mountain and Taining Ancient Town. Taining Global Geopark has a typical youth Danxia landform as the main attraction, as well as volcanic rocks, granite, structural landforms and other geological relics.

Therefore, we recommend you to first visit Taining Geology Museum at the beginning of your tour of Taining. The museum is located 2 kilometers away from Taining Town. It has miniature models of all attractions in Taining accompanied by detailed introductory texts. After visiting the museum, you will get a comprehensive understanding of the overview, background of formation, and typical Danxia landscapes in Taining Global Geopark. Once you have an idea of the beautiful natural scenery that Taining is about to present to you, you are ready to go.

And you'd better come to Taining with a poetic mood, which will enhance your experience viewing the gorgeous scenery of Danxia. Danxia is a landform that is born for the romantic aesthetics.

寨下大峡谷——一部“地质天书”

寨下大峡谷是泰宁的第一条地质科考路线，被联合国专家称为“世界地质公园的榜样景区”，位于泰宁西面的杉城镇寨下村，离县城15千米左右。

泰宁的地名里，“岩”多，如李家岩、宝盖岩、读书岩、丰岩、状元岩、栖真岩……；“寨”多，如黄石寨、盘龙寨、虎头寨、南石寨、大石寨……这些或是土匪叛乱据险，或是乡民自保避祸，都是历史上真实存在过的山寨。寨下大峡谷得名缘于所在地属于寨下村，而寨下村得名缘于村边丹霞山上曾有过大石寨、小石寨两个山寨。

寨下村是个静谧的小山村，村舍三三两两，村子呈葫芦状，背靠南面巨大的驮象岩，依傍S形的石塘溪。溪口有片风水林，古木参天。溪上有三座古桥，石面苍苔斑驳，枯藤老树相互掩映。过去这里流传着“一寺（指金龟寺）、二寨（大石寨、小石寨）、三石桥，寨下生活真逍遥”的俗话，大抵概括了山村生活哲学里的理想意境。

从村口沿着苔迹斑斑的小路绕过山弯，越过古木苍苍的天然屏障，就到大峡谷了。

◎ 泰宁世界地质公园。
Taining Global Geopark.

◎ 寨下大峡谷景区入口处。
The entrance to Zhaixia Grand Canyon.

Zhaixia Grand Canyon, a "Geological Book from Heaven"

Zhaixia Grand Canyon is Taining's first geological research route, which is recognized by experts from the United Nations as the "model of world geoparks". It is located in Zhaixia Village, Shancheng Township in the west of Taining. It is about 15 kilometers from the county town.

Many places in Taining have names ending with Yan (Rock) or Zhai (Fortress). These are all real fortresses that have existed in history and established by either bandits or the villagers for defence. The name Zhaixia Grand Canyon could be traced to the fact that it was located in Zhaixia Fortress, which had been named so because it was at the foot of a Danxia mountain with two Zhai (Big-Stone Zhai and Small-Stone Zhai) on the top.

Zhaixia Fortress is now a quiet and small mountain-village with cottages in groups of two or three. The village is in the shape of a gourd, with its back to Tuoxiang Rock in the south. It lies by the S-shaped Shitang Stream. There is a forest with towering ancient trees at the beginning of the stream and there are three ancient bridges over the stream, whose stone surfaces are covered by moss. In the forest, the withered vines and old trees cover and overshadow one another. In the past, there was a saying that "With one temple (referring to the Golden Turtle Temple), two Zhai (Dashi Zhai and Xiaoshi Zhai), and three stone bridges, what a leisurely life in Zhaixia Fortress". It summarizes the ideal artistic conception in the mountain village life philosophy.

After departing from the entrance of the village, walk along the moss-covered path bypassing the winding mountain trail and go across the natural barrier of old woods, you will arrive at Zhaixia Grand Canyon.

寨下大峡谷开辟为景区后，整条峡谷被分为三段，按地质形态的类别，被分别命名为悬天峡、通天峡和倚天峡。这三段峡谷生动地演绎了以流水侵蚀、重力崩塌、构造运动为主的三种地质现象，有三个成语可以很形象地概括这三种地质现象，即“水滴石穿”“山崩地裂”“沧海桑田”。三段峡谷首尾相连，呈环状三角形，好似一条金色苍龙蜷卧在群山之中，所以这条峡谷也被叫作金龙谷。

Zhaixia Grand Canyon is divided into three sections according to the types of geological morphology including Xuantian Gorge, Tongtian Gorge, and Yitian Gorge for its better organization as a tourist site. These three gorges vividly present three geological phenomena, which are mainly characterized by water erosion, gravity collapse and tectonic movement. The three geological phenomena can also be summarized by three respective idioms, "Constant dropping wears the stone", "The mountains explode and the earth splits", and "Seas change into mulberry fields and mulberry fields into seas". The three gorges are connected end to end in a triangular shape, like a golden dragon crouching among the mountains. This is why the canyon is also called the Golden Dragon Canyon.

◎ 三段峡谷示意图。
The Instructional Map of the Three Stretches of the Canyon.

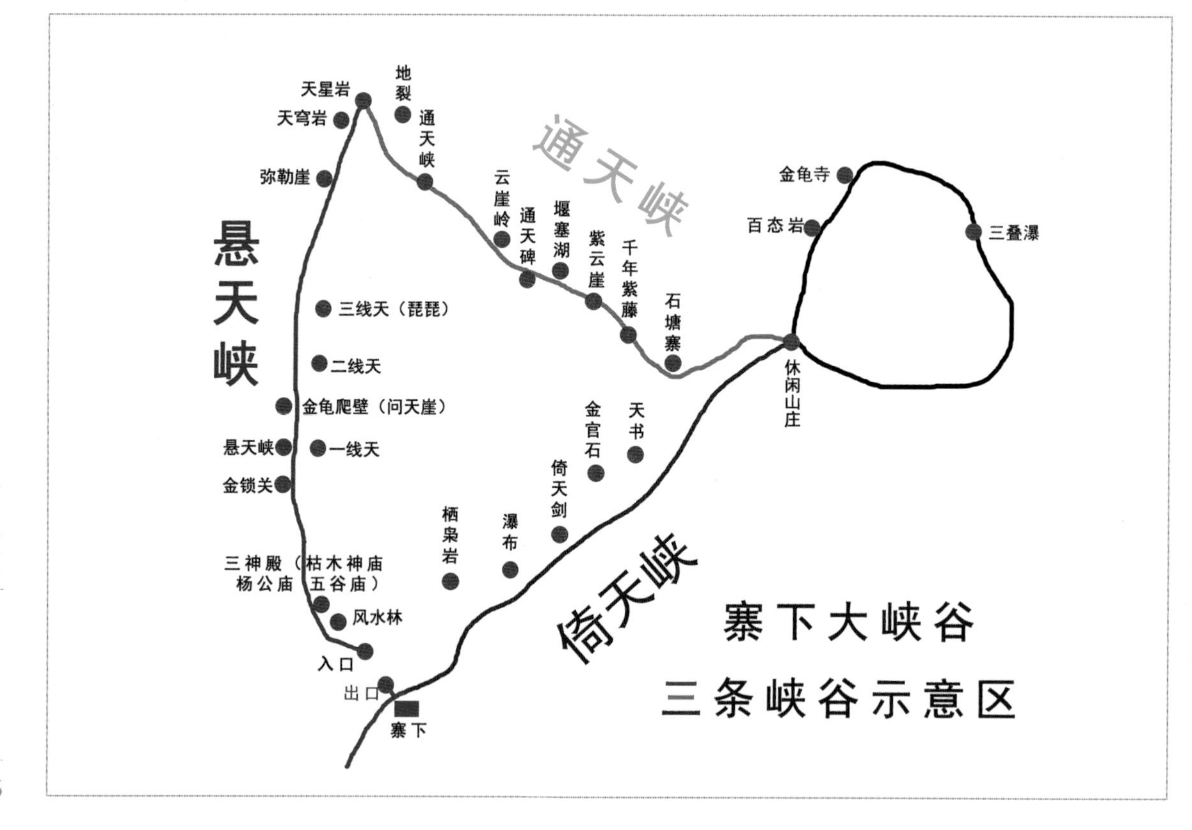

天穹岩是悬天峡段内的典型景观。在倒悬的崖壁顶端，大自然如鬼斧神工般雕凿出数百个丹霞洞穴。这是砂砾岩经流水侵蚀、风化而剥蚀形成的大型蜂窝状洞穴群。大大小小的洞穴，宛如天上的星星，布满整个崖壁。

◎ 天穹岩。
Sky-Dome Rock.

The Sky-Dome Rock is a typical landscape in the Xuantian Gorge section. At the top of the overhanging cliff, there are hundreds of Danxia caves carved by the incredible power of nature. The big beehive-shaped cave groups are formed because of the runoff erosion and weathering denudation of sandy conglomerates. The caves are all over the whole cliff, looking like stars in the sky.

◎ 悬天峡。
Xuantian Gorge.

通天峡内，因块状山体崩塌，岩石堆积谷底，形成云崖岭。云崖岭前，巨崖直立，整个崖壁就像被砂纸打磨过一样，像块高耸的碑石直插天空，所以被称作通天碑。通天碑侧面还有一块巨崖，仰望时如彩霞映照，故称映霞壁。整面崖壁的垂直高度有100米左右，长约500米，平整如墙，被当地人戏称为世界上最大的“土墙”。

◎ 云崖岭上通天碑。

Tongtian Monument on Yunya Ridge.

Yunya Ridge was formed from rocks piling up at the bottom of the valley that had been resulted from the collapse of the massive mountains inside Tongtian Gorge. In front of Yunya Ridge, there stands a straight cliff, which seems to be polished by sandpaper and looks like a towering monument piercing the sky, hence the name Tongtian Monument. There is another huge cliff beside Tongtian Monument. It looks like roseate clouds when you look up, and this is why it is called Yingxia Wall (meaning a wall that reflects the roseate clouds). The vertical height of the entire cliff is about 100 meters and the length is about 500 meters. The surface of the wall is very flat, and the locals jokingly call it the largest "earth wall" in the world.

◎ 世界上最大的“土墙”映霞壁。
Yingxia Wall, the largest "earth wall" in the world.

倚天峡清幽沉寂，泉流不息。在这里，你可以站在两个相距遥远的地质年代的分界线上，左手触摸4亿年前的古生代的变质岩，右手触摸6500万年前的中生代的沉积岩。当然，如果你对这种地质时空的激荡无感也正常，也许你会问：“它们不都是石头吗？”

◎ 倚天剑。
Yitian "Sword".

Yitian Gorge is quiet and beautiful with streams running endlessness. You can stand here at the watershed of two distant geologic ages where your left hand touches the 400-million-year-old metamorphic rocks of the Paleozoic, while your right hand is touching the 65-million-year-old Mesozoic sedimentary rocks. Of course, if you have no special feeling about the geological time and space, you might ask, "Aren't they all rocks?".

◎ 倚天峡小瀑布。

A water fall in Yitian Gorge.

上清溪、九龙潭，都是忘归处

道教有“三清三境三宝”之说，“三清”为“玉清、上清、太清”，其中“上清”象征“混沌始判，阴阳初分”的“太极”状态。上清溪的命名者，应该就是觉得这里超然古拙，如天地初始一般吧。而这里也的确与“道”有关，在上清溪下游，也就是现在竹筏上岸的地方，有个栖真岩，是道家历史上有名的梅福真人炼丹修行之地。

从泰宁古代县志、游记等文字记录来看，在泰宁山水里，上清溪一直是文人墨客心中最向往之地。明崇祯十年（1637），收到泰宁县令袁世芳的邀请，中左所（今厦门）文人池显方到泰宁玩，留下了史上关于上清溪的第一篇游记。其中游记很长，流传最广的几句是：“转一景，如闭一户焉；想一景，如翻一梦焉；会一景，如绎一封焉；覆一景，如逢一故人焉。出口几不复识，如渔夫之别桃源也。兹奇果逾九曲，以僻处无人知者。”

Shangqing Stream and Jiulong Pond, Places Where People Linger

There is a saying in Taoism, "Three Qing, three Jing and three Bao", namely Yuqing, Shangqing and Taiqing, among which Shangqing symbolizes the Taichi state of the beginning of Yin and Yang. The people who gave the name Shangqing to the stream must have thought that the environment around the stream had an ancient and transcending feeling just like the beginning of the earth. The place indeed has connections to Taoism. There is a Qizhen Rock at the place where people get off the bamboo rafts at the lower reach of Shangqing Stream. This location was where the immortal Mei Fu, a famous Taoist in the history, practiced alchemy and chanting.

According to the county records and traveling notes, Shangqing Stream was always the most longing place among literati when they came to Taining. In the tenth year (1637) of Chongzhen Period in the Ming Dynasty, the Taining County magistrate, Yuan Shifang invited the literatus Chi Xianfang from today's Xiamen to visit Taining. Chi came and wrote the first travelogue about Shangqing Stream in the history. The travelogue is very long, but the most widespread sentences reveal the infinite variety of the scenery of Shangqing Stream. The gist of these sentences is that in Shangqing Stream, you can see a different scenery with each step, which will make you feel as if you are in a dream and feel as if you are meeting an old friend. The author was fascinated about the place and did not want to leave. He found that there were indeed more than nine curved turning points here and it was so remote that it was hardly known.

◎ 百褶峡。
Baizhe Canyon.

■ 上清溪

上清溪距离县城22千米，游览时需要乘坐漂流的竹筏。这和武夷山漂流的大竹筏不同，是仅能容纳6个人的小竹筏，基本能随到随走，全程15千米，历时两小时。溪的最宽处不过10米，窄的地方仅一筏能通过，而且水流量不大，水势平缓，激流成滩处，也不过叮咚浅唱，更多的地方是凝滞成潭，如镜如梦。

武夷山九曲溪有“九曲”，但上清溪的“曲”，你是数不过来的，拐来拐去，过了一重又一重，直到你放弃努力，心甘情愿迷失在这里，随它漂流下去。

上清溪景区和九龙潭、红石沟、状元岩等景区毗邻。自县城前往这几个景点都可先坐专线车抵达长兴旅游服务区，在此购买各自景点的门票后，再转乘去各自景点的旅游车前往。

◎ 水上奇峡。
Amazing canyons along Shangqing Stream.

Shangqing Stream

Shangqing Stream is 22 kilometers away from the county town, and visitors need to take bamboo rafts for touring. These bamboo rafts are much smaller than those in Mount Wuyi, and can only accommodate six people. Rafts are available whenever visitors arrive at the dock. The whole journey is 15 kilometers and takes two hours. The widest part of the stream is no more than 10 meters, and the narrowest place can only pass one raft. The stream water flows calmly, only at some places the water runs rapidly but makes some soft sounds. While more often, the water flows together and forms a pond with a mirror-like surface. The scenery here is rather dreamy.

The Nine-Bend Stream at Mount Wuyi flows through nine meanders. However, you will not be able to count the meanders of Shangqing Stream because they come one after another and turn back and forth, until you give up on any attempt of counting, feeling perfectly happy to be lost, and continue to flow along.

Shangqing Stream tourist area is adjacent to Jiulong (Nine Dragon) Pond, Red Stone Ravine, and Zhuangyuan Rock. Tourists can take the special line to Changxing Tourist Service Area from the county town and get tickets for the attractions, and then transfer to the vehicles heading for their respective destinations.

◎ 上清溪漂流。
Drifting on Shangqing Stream.

■ 九龙潭

九龙潭景区是中国组成密度和复杂度最高的丹霞峡谷群。

潭的形成源于亿万年前大自然造就的堰塞湖，因为有九条蜿蜒如龙的山涧溪水注入潭中，潭后来就叫九龙潭。按照更地理学的表述，九龙潭是由5千米的深切峡谷和深邃的水上线谷、巷谷汇水组成的一泓清潭，最宽处100余米，最窄处不足2米，融合湖、溪、山、谷、岩、峰、沟多种形态，号称“泰宁丹霞山水微缩明珠”。

游览九龙潭景区，可以泛舟潭中，也可沿着栈道行走。栈道固定在悬崖边，沿着湖面蜿蜒，全长3,200米。其中，2017年新增了一段玻璃栈道，长60余米，垂直高度约30米，位于虬龙峡与晴龙峡两个峡谷之间。自从河北白石山因玻璃栈道而游客爆满之后，全国各地景区的玻璃栈道如雨后春笋涌现，但多以“险绝”作为卖点，游客体验的是心惊肉跳的感觉。但九龙潭的玻璃栈道却是为了更好地衬托这里“幽谷水巷”的秀美。

◎ 泛舟九龙潭。
Drafting down Jiulong (Nine Dragon) Pond.

◎ 幽谷水巷。
Quiet water alleys.

Jiulong Pond

The Jiulong (Nine Dragon) Pond is among the Danxia Gorge Group with the highest density and complexity in China.

The formation of the pond originated from the barrier lakes that had been created by nature hundreds of millions of years ago. The reason why it was later called Nine-Dragon Pond was that there were nine mountain streams meandering like dragons flowing into the pond. According to geographical explanation, Nine-Dragon Pond is composed of the merging water from a five-kilometre-deep canyon, deep V-shaped valleys and U-shaped valleys. The largest width of the pond is more than 100 meters and the narrowest is less than 2 meters. A variety of Danxia landforms such as lakes, streams, mountains, valleys, rocks and peaks make "the miniature pearl of Taining Danxia landscape".

When visiting the Nine-Dragon Pond scenic area, you can go rafting on the lake, or walk along the plank path. The walkway is fixed at the cliff edge and meanders along the pond. The total length of the walkway is 3,200 meters. In 2017, a new section of glass walkway was added, which was more than 60 meters long and about 30 meters in vertical height. Glass walkways have sprung up in tourist spots across the country since Hebei's White-Stone Mountain was flooded with tourists. Most of the glass walkways adopt the idea of "dangerous and unique" as their selling points. While the glass walkway at Nine-Dragon Pond is to set off the beauty of the quiet water alleys.

◎ 溪谷秋韵。

The beautiful scenery of Danxia valleys in autumn.

不论是在上清溪还是在九龙潭，人在筏中坐，筏在水中走，身处这些收束在深山峡谷之中的溪流深潭里，感受最深的就是空灵幽静。

“芳树无人花自落，青山一路鸟空啼。”此种情景里，最适合玩一个心灵游戏，就是默语凝神，去想眼前的一切亘古如斯。静水，怪石，崖壁上断线的水帘，袭来幽香的一丛兰花，嫣然绽放在石壁缝隙里的一朵杜鹃，还有水汽氤氲处薄薄的一片苔藓，甚至是崖上一瞥掠过的山鹰，不知道哪个角落发出的某种山兽若有若无的一声低嗥，这些不都是打破了时间的魔咒，恒久不变的吗？

那我们此刻又处在什么样的时间，是古，还是今？时间到底是什么？什么又才是生命的存在？这些无用之问，是在这方山灵的启迪下，有心人才可以打开的人生安慰剂。

山外之人，从疲于奔命的世俗生活中赶来，不就是希望把自己疲惫的身体和精神放置于这样一方纯净通灵的山水之间，得到休憩，并找到新的力量吗？

而两岸岩壑深邃，或幽或明，峰回路转，步移景换，却也如人生旅途之况味。一路下来，可不发一语，默默感悟，或者只是独自浅吟。实在心绪难发，倒是也可以来一声发自肺腑的长啸，于是空谷传响，瞌睡的鸟惊得飞起，扑棱扑棱地地扇动翅膀，扯破静止的空气，终于打断了你的沉思。回过神来，天已日暮，沉醉不知归路。

最完美的还是要邂逅一场小雨，或细细蒙蒙下不停，或噼里啪啦来一阵，总会助心境再美上一层，“山头斜照却相迎。回首向来萧瑟处，归去，也无风雨也无晴。”

Whether at Shangqing Stream or at Nine-Dragon Pond, when people sit on the rafts, drifting on the water, immersing themselves in deep canyons and valleys, their deepest feelings are always ethereal and quiet.

"The flowers fall though no one notices, the green mountains are so empty all that can be heard is birds singing." There is a meditative game that is suitable in this kind of environment, and it is to stay in silence and focus on everything around: Still water, strange rocks, intermittent waterfalls on the cliffs, a tuft of fragrant orchids, a rhododendron that blooms in the gap between the stone walls, a thin layer of moss on the ground under dense water vapor, and even the eagle that skims over the mountains, the howl of the unknown mountain animals... don't they all break the spell of time, and become timeless?

Then which time zone are we in at the moment, ancient or present? What exactly is time? What is the existence of life? These useless questions are the life placebo that can only be opened by people with heart under the enlightenment of these mountains.

Visitors come here, escaping from the exhausting mundane things of life. Don't they all wish to indulge themselves, both tired body and spirit, in such a pure environment to have a rest, recharge and rejuvenate?

The diverse scenes of the two sides are just like those of the journey of life—the deep rocks can sometimes be dark, sometimes bright; the path winds along mountain ridges; with each of your step you get a different view. All the way down, you can keep quiet and reflect in your heart, or you can chant in peace. If you have much accumulated emotions and have held them within, you can give a long shout from the bottom of your heart, which may echo through the empty valley and wake up the sleepy birds. The sound that these birds make when they flap their wings will break the peace in the air, and finally interrupt your deep meditative state.

Before you realize, it is already dusk. The scenery is so charming that makes you intoxicated and forget the way back.

It would be perfect if you encounter some light rain, either drizzles for a while or pours suddenly, it always adds another layer of serenity to your mood. Like the poem describes, "The sunset at the top of the hill greets me. Looking back at the place where I encountered the wind and rain, I felt calm. The way back now will have no stormy rain nor sunshine."

八仙崖——东南海拔最高的丹霞山峰

泰宁有一个地方，很多来这里旅游的人应该都遥望过它的身姿，因为它的主峰大牙顶海拔907.6米，是大金湖丹霞地貌的最高出露区，也是中国东南部丹霞地貌的最高峰。在环大金湖旅游区域的很多著名景点都可以眺望到它，但身临其境的人并不多。

这个地方就是八仙崖，在泰宁南部的大龙乡。大龙乡算是泰宁最偏远的乡镇，过去道路崎岖险阻。这里的风景有别于早已名扬四海的大金湖、寨下大峡谷、九龙潭、上清溪等，一直鲜为人知。

八仙崖自主峰大牙顶向西南1千米范围内依次有大牙顶、脚趾石、东门牙石、西门牙石、东圣教石、西圣教石、菜刀石及未名石8座800米以上的峰柱，以发育典型的丹崖、岩峰、岩柱地貌为特色，雄伟壮观，云雾缥缈时仿若仙山。有诗这样描写八仙崖："仙石婷婷屹八尊，红云缭绕紫烟含"，很是形象。

老鹰在蓝色苍穹之下，掠过层云掩映的丹霞崖壁，呼唤造访此地的人。这些年，慢慢有不少户外俱乐部或者资深驴友登八仙崖，并在此露营、野餐、看云海、看日落日出，这些成为泰宁深度游的新形式。

Baxianya, the Danxia Peaks with the Highest Altitude in Southeastern China

There is a place in Taining that many tourists have seen it from a distance, but few have been to. Its main peak, Daya Peak, is 907.6 meters above the sea level, which is the highest exposed area of Danxia landform at the Dajin Lake area. It is also the highest Danxia peak in southeastern China, which can be seen from many famous scenic spots in the tourist area of Dajin Lake.

That is Baxianya (or Eight-Immortal Peaks) located in Dalong Township in the south of Taining. Dalong Township is the most remote one in Taining and very hard to reach. The scenery here is different from the well-known Dajin Lake, Zhaixia Grand Canyon, Nine-Dragon Pond, Shangqing Stream and others. It is rarely known by people.

◎ 中国东南海拔最高的丹霞山峰八仙崖。
The Danxia peaks with the highest altitude in southeastern China—the Eight-Immortal Peaks.

At Baxianya, within the range of 1 kilometer to the southwest of Daya Peak lie 8 peak columns with a height over 800 meters. They are characterized by the typical well-developed cliffs, peaks and rock pillars. When they are twined by clouds, they look like eight immortals, hence the name Baxian (eight immortals). There is a poem that vividly depicts the peaks, which says, "The eight immortals stand there with dignity, covered by red clouds and twined by purple mist."

Under the blue sky, the eagles fly over the cloud-covered cliffs of the Eight-Immortal Peaks, calling visitors to come. Over the years, there have been many outdoor club members or senior tourists that have climbed the peaks. They camp, have picnic, watch the clouds and the sunset and sunrise, these have become new trends of in-depth travel in Taining.

八仙崖脚下有个深山小村落叫张地村。按照《张氏族谱》所写，当地村民是唐朝宰相张九龄的后人。明宣德年间（1426—1435），张九龄第28代孙张洪由建宁迁至泰宁龙安时，途经八仙崖，“见其山水秀丽，筑室而居”。张洪之后，又历23代至今。

张九龄，是唐朝开元年间（713—741）名相、诗人。他的那首经典的《望月怀古》流传至今。如果你哪一天在八仙崖露营过夜，恰逢有明月的话，倒是可以尝试着风雅一番。在山顶对着云海吟诵此诗，会是一种绝佳的体验。“海上生明月，天涯共此时。情人怨遥夜，竟夕起相思。灭烛怜光满，披衣觉露滋。不堪盈手赠，还寝梦佳期。”其意境雄浑开阔，又幽邃细微，对此地此景最适合不过了。

◎ 俯瞰张地村。

A bird's eye view of Zhangdi Village.

There is a small village called Zhangdi Village at the foot of Baxianya. According to the *Genealogy of the Zhang's Clan*, the local villagers are descendants of Zhang Jiuling, the prime minister of the Tang Dynasty. Between 1426 and 1435, Zhang Hong, the 28th generation desendant of Zhang Jiuling, moved from Jianning to Long'an in Taining. When he passed the Baxianya, "he found it beautiful and peaceful, so he built his house and lived there". There has been 23 more generations since then.

Zhang Jiuling was a famous prime minister and poet of the Tang Dynasty during 713—741. His classical poem *Remembering the Past While Looking at the Moon* has passed on until today. The poem is suitable for reciting on a night you camp on Baxianya with a bright moon—to stand on the top of the mountain, facing the sea of clouds and chanting the poem. The poem has a powerful, open, deep and quiet conception, which is exactly the depict of the moment. Here is how the poem goes:

◎ 仙石屹立。
Like an immortal standing with dignity.

As the bright moon shines over the sea,
From far away you share this moment with me.
For parted lovers lonely nights are the worst to be.
All night long I think of no one but thee.
To enjoy the moon I blow out the candle stick.
Please put on your nightgown for the dew is thick.
I try to offer you the moonlight so hard to pick,
Hoping a reunion in my dream will come quick.

(Translated by Ying Sun)

◎ 藏棺岩。
Cangguan Rock (the Rock for Hiding Coffins).

值得一提的还有，张地村最出名的是其相当规模的藏棺数量。行至张地村附近一个叫紫云岩的地方，就可以看到岩穴中一排排码放整齐的棺材。村里的习俗是，自男丁出世到三四岁之间，家里人就得为他做好一具棺材放在洞穴中藏匿，代代相传的说法是，这样可以使人长寿。等他去世后，家人再取出棺材，将其安葬到别处。这些年实行火葬后，这种藏棺的民俗活动也停止了。目前紫云岩里的棺材数量还有近百具，只是作为一种遗存留给人参观及供专家们研究了。

It is worth mentioning that Zhangdi Village is best known for its considerable number of the hidden coffins. There is a place called Ziyun (Purple-Cloud) Rock near Zhangdi Village, where you can see rows of coffins neatly placed in the caves. The custom in the village is that from the time a boy is born to the age of three or four, his family will make a coffin for him and hide it in a cave. It is said that this can help with longevity. After his death, the family takes the coffin out and bury it elsewhere. Since cremation has been practiced in recent years, the custom of hiding coffins stopped. Currently, there are nearly 100 coffins in Ziyun Rock, which are left as relics for visitors to see and for experts to study.

◎ 仙人聚首。

Immortals' gathering together.

◎ 八仙崖下古村落。

An ancient village at the foot of the Eight-Immortal Peaks.

小城营造——尚书第到明清园

泰宁是一个人杰地灵的小城。无论宋代还是现在，它的县城人口都不到14万人。就这样的一个小城，历经一千多年，生生不息，并且还能谱写出“隔河两状元，一门四进士，一巷九举人”的人文鼎盛，不能不算是一个小小的奇迹。这里藏着先人的智慧，也藏着不尽的故事传说。它们和那些城市营造变迁中留下的看得见的印记——街巷、民居、城墙、桥梁、古井、牌坊等，一起交叠印证，成为你去领略其魅力的又一把钥匙。

The Constructions in Taining: From Shangshu Mansion to Mingqing Palace

Taining is such a great place. Either in the Song Dynasty or now, the population here has never surpassed 140,000 over a long history of more than a thousand years. The small town is a wonder and has had its humanistic phenomenon. There were two champions of ancient imperial examinations who lived on the opposite sides of the river, four Jinshi (the third-degree scholars in the highest imperial examination) from the same family and nine Juren (a successful candidate in the imperial examinations at the provincial level) from the same lane. Here, the wisdom of the ancestors, the endless stories and legends, together with those visible marks left by the transformation of the town—streets, dwellings, walls, bridges, ancient wells, archways, form a key that guides people to explore the charms of Taining.

◎ 雨中尚书巷。
Shangshu Lane in the rain.

千年泰宁城

泰宁最早建县是在958年的南唐，当时叫归化县，城建布局还没有什么章法。一直到了宋朝，人口增多，城市规划迫在眉睫。于是，泰宁人重金聘请了云南东川的范越凤来做县城的规划，那时候管这项工作叫“相度”。

范越凤到了泰宁，跋山涉水，看了山脉形势和溪流走向，然后提出把城址向东移500多米，将县衙和民居“始迁于庐[炉]峰之下，杉水之南”，形成青山抱城、溪水绕城的泰宁城的基本格局。找个地方登高俯瞰泰宁城，可以看到“金溪水环绕山城，半壁合抱弯如长月，古城浑若红日坐落当中”。

确定城址位置之后，人们开始布局水道，开挖水井，然后在北溪拦河筑坝，把河水引入下水道，既解决了居民日常用水，又发挥了消防功能，还能分洪排涝。

据说重建后的泰宁，万象更新，物阜民丰，而且人才辈出，人文鼎盛。

1070年，宋代，叶祖洽成为泰宁史上第一个状元；1196年，邹应龙也考取状元。整个宋代，泰宁共出了23名进士，这种盛况在之后却再也没有出现过：元朝，泰宁没有进士，只有几位举人；明朝，泰宁有进士7人，举人24人；清朝，进士只有1人，举人29人。

◎ 状元雕塑。

A statue of the champions of imperial examinations in the Song Dynasty.

Thousand-Year-Old Taining Town

Taining County was first established in 958 AD during the Southern Tang Dynasty. It was called Guihua County at that time and the layout of the county was arbitrary. With the increased population during the Song Dynasty, further planning of the county became a pressing issue. As a result, Taining people paid a large sum of money to hire Fan Yuefeng from Dongchuan of Yunnan Province, to do the planning of the countytown.

Upon Fan Yuefeng's arrival in Taining, he climbed mountains and walked along streams to check the patterns of the mountains and the directions of the flows of the streams. He proposed to move the town to the east by more than 500 meters, and to move the county government and residential houses "to the foot of the Lu Peak and to the south of Shan Stream", to form the basic layout of Taining Town, where it is surrounded by green mountains and encircled by streams. If you find a place to have an overview of Taining, you will see a scene where "Jinxi Stream surrounds the town, and the mountains around are curved like the crescent moon, and the ancient town sits in the middle like a red sun".

After the location of the town was determined, people began to design waterways and dig wells. They then dammed the Beixi River and designed drainage, which not only solved the daily water use of the residents, but also served well in firefighting as well as diverting flood.

After re-planning the layout of the town, it was said that everything of Taining took on a complete new look—produces were plentiful and people became wealthier. Since then, a large number of talents from Taining stood out and Taining experienced a period of flourishing humanistic environment.

In 1070 of the Song Dynasty, Ye Zuqia became the first Zhuangyuan—the champion of the highest imperial examination in the history of Taining. In 1196, Zou Yinglong also ranked the first in an imperial examination. Throughout the Song Dynasty, Taining had a total of 23 Jinshi—the third highest ranked scholars in the imperial examination. Such remarkable period of human talent of the Song never happened again in this area. In the Yuan Dynasty, Taining did not have any Jinshi, but only several Juren—a successful candidate in the imperial examinations at the provincial level. There were 7 Jinshi and 24 Juren in the Ming Dynasty; and only 1 Jinshi and 29 Juren in the Qing Dynasty.

◎ 雕花基座。
Carved pedestals.

只是要真的看清这种科举不兴，是要把泰宁放置在更大的历史视野中的。泰宁籍的作家萧春雷在文章《泰宁科举外史》中作过分析。

宋代福建科举大盛，全国共计进士28933名，其中福建占了7144名，全国第一。而宋代，也是泰宁科举的全盛时期。明清时期，福建文化以及科举的领先地位已经被江浙取代，与之相应，泰宁的科举水平也在衰败，并且衰败的速度要更快。

但是我们还可以这样理解，泰宁"僻在万山之中，弹丸黑子"，人口不过十万的山区小县，能有这样的成绩，可以说已经很了不起了。

走在今天的泰宁古城，也许你还会在街边看到残存的石旗杆、雕花基座等。这些功名坊表留下碎片记忆，带着泰宁城过去的荣耀，引发你关于这个地方的思古之幽情。

It is necessary to put Taining in a larger historical perspective if we really want to have a good idea of its destiny. Xiao Chunlei, a writer from Taining, has offered an analysis in his article "*The History of Taining's Imperial Examination*".

In the Song dynasty, Fujian experienced a great success in the imperial examinations. Fujian had 7,144 in a total of 28,933 Jinshi scholars of the nation, ranking first in the nation. The Song Dynasty was also the heyday of Taining in the imperial examinations.

During the Ming and Qing dynasties, Taining was surpassed by Jiangsu and Zhejiang provinces in the imperial examinations. Thus, the performance of Taining scholars in the imperial examinations was also declining, and the speed of decline was faster. But as a small mountain town with a population of no more than 100,000, Taining's achievements were quite remarkable.

Walking in today's ancient town of Taining, you may still see remaining relics like stone flagpoles and carved pedestals. These relics contain a lot of memories and show the glory of the county in the past, triggering your nostalgia about this place.

◎ 旗杆石。
Stone flagpoles.

◎ 大东门井。
Dadongmen Well .

水井、城墙、广场

“有水井处有人家”，一个城市的形成，先从村到圩市，再到城镇，最早或许就是从最初几户人家聚居而后建的一口井开始。《周易·井》称“改邑不改井”，水井一旦开掘形成，其生命力有时候比城邑更持久。

古井是文物，因为它岁月悠长；但它又不是文物，因为它还是那样新鲜。在今天泰宁的巷子里，很多人一天的生活还是习惯从古水井的一担清水开始；水井周边甚至还是街坊日常生活中最重要的“民间议事厅”，多少家长里短、八卦传奇，都在水井边有滋有味地流传。今天，若你在泰宁县城问路，当地人还是会习惯地以古井作为方位参照物，先告诉你某个古井的名称及位置，再指出去目的地行走的路线。

泰宁城区至今保存了从唐代到明清各个时期的水井，各水井的井沿外壁镌刻着井名和建井人的姓名及建井日期，还刻有一些修饰图案。水井有以宗教信仰命名的，如圣公井、天王井、土地堂井；有以家族为纽带命名的，如毛家井、卢家巷井；还有表示尊崇儒家思想的水井名，如儒学井、兴贤井、崇仁三井等。如果将这些古井按历史年份排列，呈现在我们面前的就不再是一口口孤立的古井，而是纪录泰宁变迁及不同历史阶段人文特征的履历表。

比如藏在九举巷里的“崇仁三井”。传说元末明初，泰宁的何恩夫妇久未生育。何恩见城东一带的居民靠挑河水饮用，但每年春天，阴雨连绵，河水浑浊，不能饮用，便出钱雇人开了三口水井，即崇仁三井。何恩夫妇后来生了个大富大贵的儿子何道旻，以闽中选贡第一名入国子监，为官40余年，宦迹遍及八省。明太祖朱元璋颁诏褒扬他“器度尊严，才猷卓拔”，说他是有名的“何青天”。崇仁三井现仅存两井，即牌楼下井和大井头井，另一口大东门井于1989年建尚书巷时被填平，井圈现放在风水轮旁供游人参观。

Wells, Walls, Squares

The formation of a city sometimes may start from a well built by a few families living together. Then more people came and formed a small village, a town, and finally a city. As the old saying goes, "Where there is a well, there are families." A well has a very long life-span. Once a well is dug, it sometimes outlives a town. The piece about wells in *The Book of Changes* also says, "Even the city changes, the well does not."

Ancient wells are relics because they have long history; but they may not be relics since they are still fresh. Today, many people living in the alleys of Taining still have a habit of starting their day with getting two buckets of fresh water from the ancient wells. The wells are even the most important "private congress hall" in the daily life of the neighborhood. People talk about their life and gossip beside the wells fervidly. Nowadays, if you ask for directions in Taining Town, the locals will still habitually use the ancient wells as reference points. They will first tell you the name and location of an ancient well and then point out the route of the destination.

The wells in Taining Town from the Tang Dynasty to Ming and Qing dynasties are well preserved. The names of these wells, the names of their builders and the dates of their construction as well as some decorative patterns are engraved along the outer wall of wells. Some of the wells are named after religious beliefs, such as Shenggong (Saint) Well, Tianwang (God) Well, and God of Earth's Well; some are named after the family names, such as Mao's Well, Lu's-Alley Well; and some of the names also show reverence for Confucianism, such as Confucian Well, Xingxian (Virtue) Well, the Three Chongren (Benevolence) Wells and many others. If we line these ancient wells up according to the historical order, they are no longer a series of isolated ancient wells, but a record of the vicissitudes of Taining and its humanistic characteristics at different historical phases.

◎ 牌楼下井。
Pailouxia Well .

There is a story about the three wells called the Three Benevolence Wells that are hidden in the Jiuju Alley. It is said that at the end of the Yuan Dynasty and the beginning of the Ming Dynasty, there was a married couple that could not conceive for a long time. He En, the husband, sponsored the digging of the three wells, namely Benevolence Wells, for villagers who had lived in the east of the town. He En did so because he had seen that the continuous rain in spring had turned the river water undrinkable for the villagers that all had survived based on the water supply of the river. He En and his wife giving birth to a healthy baby boy called He Daomin. Daomin later entered the Imperial Academy as the champion of Fujian Provincial examination and served as an official for over 40 years overseeing eight provinces at different points of his tenure. Zhu Yuanzhang, the Emperor Taizu of the Ming Dynasty, issued an imperial edict praising his dignity and talent and granted him the title of "Qingtian (Justice) He". Only two wells of the Three Benevolence Wells have remained until today, which are Pailouxia Well and Dajingtou Well. The Dadongmen Well was filled during the construction of Shangshu Lane in 1989, the ring of the well was placed next to the Water Wheel for visitors to admire.

◎ 大井头井。
Dajingtou Well.

◎ 儒学井（又名红军井）。
Confucian Well (also called the Red Army Well).

再比如岭上街的红军井，原来叫儒学井，20世纪30年代时，这里曾驻扎着大批红军官兵，包括周恩来、朱德等都在这里生活过。他们的生活用水都是从这口井里汲取的。为了纪念这段历史，人们将这口井改名为“红军井”。

从井的分布和命名里就可以读出泰宁过去的样子和故事。你可以漫步于不大的泰宁县城，来一趟水井寻访之旅。

Another example is the Red Army Well on Lingshang Street, which was originally called the Confucian Well. In the 1930s, a large number of Red Army officers and soldiers were stationed here, including Zhou Enlai and Zhu De. Their domestic water was taken from this well. In order to commemorate this history, people renamed this well "Red Army Well".

It is possible to depict an idea of the past scenes and stories of Tanning from the distribution and the naming of the wells. You can also stroll through the small town to have an exploration tour of water wells.

除了水井，城墙也是城市曾经最重要的见证者。

泰宁依山傍水，长期只倚重山水之险，未筑城墙，兵寇一来，百姓只能拖家带口往深山里躲。修筑城墙这事一直到明朝嘉靖年间才得以解决。

1560年，即明嘉靖三十九年，南昌举人熊鄂到泰宁当知县，“切为斯民忧之”，想尽办法筹钱修筑城墙，耗时6个月告成。修好的城墙，三面据河，西面依傍炉峰山，高5.3米，厚2.3米，周长2363米，有四扇门，东曰左圣，西曰右义，南曰泰阶，北曰朝京；又有四扇小门，东曰昼锦，西曰靖远，南曰莪仁，北曰青云。

In addition to water wells, the city wall is also the most important witness to the city.

Taining is surrounded by mountains and water, and for a long time, the city relied solely on the natural barriers of the mountains and rivers for defence purposes and did not have city walls. As a result, when the enemies came, people could only hide in the mountains with their families. The building of the rampart was not settled until the Jiajing Period (1796—1820) of the Ming Dynasty.

In 1560, the 39th year of the Jiajing Period of the Ming Dynasty, Xiong E from Nanchang was appointed the governor of Taining County. He cared a lot about his people, and did all that he could to raise money for building city walls for Taining. It took six months to complete the construction of the walls. The three sides of the walls at its completion faced rivers and the west side of the wall was attached to Lufeng Mountain. It was 5.3 meters high, 2.3 meters thick and 2,363 meters in circumference. It has four main gates and four small gates, but only one remains today.

◎ 泰宁古城墙。
The ancient city wall of Taining Town.

之后城墙经历多次水灾、匪乱，多次修复、重建，至20世纪50年代末拆除，再后来因城市扩张，只留下一小段古城墙，残留的城垣中有一扇小门，即昼锦门。

《史记·项羽本纪》曰："项王见秦宫室皆以烧残破，又心怀思欲东归，曰：'富贵不归故乡，如衣绣夜行，谁知之者！'" "衣绣夜行"的反义词就是"衣锦还乡"，相传昼锦门就是为了表达对泰宁第一个状元叶祖洽的尊敬而命名的，寓意状元衣锦还乡。

The city walls went through many floods, disturbances, repairs and reconstructions, and were demolished in the late 1950s. Later, because of urban expansion, only a small section of the ancient city wall was left. There was a small gate in the remaining wall—that is Zhoujin Gate.

The Historical Records (Shiji) • *Xiangyu* writes, "The King Xiang saw the palace of Qin being burnt down, and wanted to head to the East back to his hometown. He thus found an excuse, 'If one becomes successful and wealthy but does not go back to his hometown, it is like a person wearing a splendid dress for a night out, who would know that he is rich?'" It was said that one of the gates was named Zhoujin to show respect to Ye Zuqia, the first champion of the imperial examination in Taining's history, implying that Ye had come back home after obtaining fortune and success.

◎ 古城墙上还刻有明朝广告："溪埠上下不许堆积，违者罚钱六百！"
A notice of the Ming Dynasty can be seen on the ancient wall, reads "No rubbish around the port; violators will be fined 600".

2005年，泰宁在古城墙遗址上，即现在东洲桥一带，沿河边建起了状元文化公园，其中千年赋青铜塑像公园里有19组青铜雕塑群，诠释了泰宁从战国末期至红军长征北上近2200来的历史人文故事。公园中沿河开阔处是状元文化广场，是今日泰宁县城最热闹的中心所在。

青铜塑像群中最醒目的雕塑是“将相对弈”。在高大的两轮风水车（寓意风水常转）旁，塑有一座高台，端坐其上的是泰宁历史上仕途最显赫的两个人物，一位是南宋状元，后任枢密院副使，参知政事（副宰相）的邹应龙；一位是明朝进士李春烨，任少保兼太子太师协理京营戎政兵部尚书。

◎ 青铜塑像公园。
The Millennium Bronze Statue Park.

In 2005, Zhuangyuan Cultural Park was built on the historical site of the ancient city wall along the riverside near Dongzhou Bridge. The 19 bronze sculpture groups in the park present the historical and humanistic stories of Taining from the end of Warring States Period to the Red Army's Long March over a period of history of more than 2,200 years. The Zhuangyuan Cultural Square in the park was along the open area of the river, which became the most vibrant center of Taining Town today.

The most striking sculpture of the park is "Chess competition between the general and the prime minister". There is a high platform next to the two tall wheeled geomantic cars (meaning good luck rotates), with the sculptures of the two most prominent characters in the history of Taining sitting on it. One of them was Zou Yinglong, the champion of an imperial examination in the Southern Song Dynasty, who later was appointed the Deputy to the Privy Council and was once a high official (deputy prime minister). The other was Li Chunye, a scholar from Taining of the Ming Dynasty, who worked for the prince and was honored the prince's teacher in addition to be the assistant to the minster of the military department.

◎ "将相对弈"。
Chess competition between Li Chunye and Zou Yinglong.

尚书第的秘密

今天指的泰宁古城区，大致就是指绣衣坊一片，保持着几个世纪形成的基本格局，青砖黛瓦，巷子深深，泰宁历代高官大多居住在这一片，建筑绝大部分坐西朝东，背靠着西炉峰山龙脉，面向城东奔腾而来的水系。

◎ 绣衣坊。
Xiuyi Lane.

这其中最精华的，就是李春烨的故居尚书第，是现存于世的中国明朝建筑艺术的瑰宝。

尚书第加上建于元明时期的世德堂，建于清朝中叶的李氏宗祠，以及梁家宅院、江家宅院、梁家巷、福堂巷等，共同构成尚书第古建筑群，总面积达12000平方米。该建筑群中有古代官居、民宅、祠堂、客厅、辅房、大庭院等多种功能建筑，保留了以明为主、包含清及民国时期至今的、不同时期的建筑风格，浓缩了完整的中国古代南方民居的生活场景，是国内保留下来的为数不多的古代民居精品建筑群落之一。

The Secrets of Shangshu Mansion

Today, the ancient town of Taining refers to the area of Xiuyi Lane. It maintains the basic layout that has formed in the past few centuries with black bricks, tiles and deep alleys. Most of the senior officials of past dynasties of Taining lived in this area, and the buildings faced the east. At the back of these buildings were the mountain range of the West Lufeng Mountain, while facing the rivers and streams coming from the east of the town.

The most essential one is Shangshu Mansion, which was the former residence of Li Chunye. It is the treasure of the remaining Chinese architectural art of the Ming Dynasty.

The ancient architectural complex consists of Shangshu Mansion, Shide Hall built in the Yuan and Ming dynasties, Li's ancestral temple built in the middle of the Qing Dynasty, Liang's House, Jiang's House, Liang's Alley and Futang Alley with a total area of 12,000 square meters. The building complex includes ancient official residences, private homes, ancestral halls, living rooms, auxiliary rooms, large courtyards and other functional buildings. It retains different architectural styles of mostly the Ming Dynasty, but also those from the Qing Dynasty to the Republic of China up until now. The complex condensed the living scenes of the complete ancient Chinese dwellings in the south, and is one of the few well preserved ancient residential architectures in China.

◎ 俯瞰尚书第建筑群落。

A bird's eye view of the ancient architecture complex around Shangshu Mansion.

在泰宁的科举史上，几位代表泰宁走向全国的人物最是耀眼：宋代是叶祖洽和邹应龙，他们也是泰宁历史上仅有的两名状元；明代是江日彩和李春烨。

叶祖洽官至吏部侍郎，历经仕途是是非非，最后客死他乡。泰宁城里，也就“昼锦门”据说是为纪念他而命名。

邹应龙官至参知政事、签书枢密院事，但方直正派、两袖清风，晚年辞官归隐泰宁南郊，今天的状元岩景区是他的读书处，算是他的一点遗迹。

江日彩，官至太仆寺少卿，他在泰宁的故居只是今天进士巷里一个普通的宅院。相比之下，兵部尚书李春烨在泰宁留下的占地5400平方米的尚书第、占地1040平方米的尚书墓，还有他读书的李家岩，就显得格外隆重。

◎ 叶祖洽。
Ye Zuqia.

◎ 邹应龙。
Zou Yinglong.

◎ 江日彩。
Jiang Ricai.

◎ 李春烨。
Li Chunye.

In the history of the imperial examination in Taining, there were several scholars that represented Taining and became well-known at the national level for their distinguished achievements: Ye Zuqia and Zou Yinglong in the Song Dynasty (who were the only two champions in the history of Taining), and Jiang Ricai and Li Chunye of the Ming Dynasty.

The highest appointment for Ye Zuqia in his career as an official was the minister of the Ministry of Personnel. He had lived a career life filled with ups and downs and died away from home. It is said that in Taining Town, only Zhoujin Gate was named to commemorate him.

Zou Yinglong was appointed the Deputy to the Privy Council. He was honest and had integrity. He resigned and lived a secluded life in the southern suburbs of Taining in his later years. The tourist site Zhuangyuan (Champion) Rock was where he read and studied.

Jiang Ricai was appointed an official of Taipu Temple (the department in charge of the transportation). His former residence in Taining is just an ordinary house in Jinshi Alley today. In contrast, the 5,400-square-meter Shangshu Mansion, 1,040-square-meter Shangshu Tomb and Li's Rock left by Li Chunye, the minister of the military department, are quite grandeur.

◎ 纪念叶祖洽而命名的昼锦门。

Zhoujin Gate named to commemorate Ye Zuqia.

◎ 江日彩故居。

Jiang Ricai's former residence.

◎ 邹应龙读书处。

The study space for Zou Yinglong.

◎ 尚书第门楼。
The gate way of Shangshu Mansion.

李春烨当官，从1624年的从七品，到1627年的从一品，至兵部尚书，加官少保兼太子少师，连升带跳11级只花了三年多，令人眩目的火箭式升官速度，至今仍是一个谜。

明朝官员工资不高，李春烨46岁才中进士，做官时间不算长，哪里来的巨额资金盖这么一大片奢华的宅院？当地有个传说，他当年运金银财宝回家，让泰宁八大城门开了三天三夜才运完。这也是一个谜。

但可以确定的是，李春烨晚年过得不开心。1629年，崇祯皇帝处理权倾一时的太监魏忠贤及其党羽，李春烨被定了一个勾结宦官的罪名，虽然后来交了罚款，不必服刑，但转眼又变为庶民。从此，李春烨就待在泰宁老家，在自己刚刚修建的尚书第里闭门养老，一直到去世。

历史真相总在一团迷雾中，但建筑却能穿越历史的迷雾，把确切的美和赞叹定格在眼前。

走进尚书第的高墙大院，门户深深，既有京城官府建筑的恢宏气势，又糅合了当地府第式建筑和徽派建筑的元素，是当时中国南北建筑工艺的结合。

Li Chunye started his career as an official from the lowest rank in 1624 to the highest rank in 1627, working as the minister of the military department. He was also assigned the additional role working for the prince as his safety guard. Li had been promoted 11 ranks in just over 3 years, and it is still a mystery how he could get promoted at such rocket speed.

The salary of the officials of the Ming Dynasty was not high and there was another mystery of Li Chunye as to why he had so much fortune to build such a luxurious home. After all, Li won the title of Jinshi only when he was 46 and his career was not long. It was said that the eight main city gates of Taining remained open for three days and three nights to allow Li to transport all of his properties home.

But it was certain that Li Chunye did not have a happy life during his old age. In 1629, when the Emperor Chongzhen convicted the eunuch Wei Zhongxian and his clique, Li Chunye was also convicted of colluding with the eunuch. Although he paid fines to be exempted from jail time, he fell from a senior official to common person. Since then, Li Chunye lived his life in retirement in his Shangshu Mansion until his death.

The truth of the history may always be mysterious like the fog, but the architecture can transcend the historical fog and show us its beauty and magnificence.

Shangshu Mansion has a high-walled courtyard and a deep structural layout from the gate to the inside. They display the grandeur of buildings in the imperial capital city, but also have blended the elements of local architecture and Hui-style buildings. It offered a combination of the techniques of Chinese construction of the North and the South at that time.

尚书第的主体建筑有五幢，坐西向东呈“一”字形排列，彼此间用封火墙相隔。尚书第后有花园，前有甬道。除了五幢主体建筑，尚书第还有书院、辅房、马房、后花园等附属建筑。除厅堂、天井、回廊外，共有120余间房，均系砖、石、木结构。

There are five main buildings in Shangshu Mansion, which are arranged in the shape of a straight line from the west to the east, separated from each other by fire walls. There is a garden behind Shangshu Mansion, and a corridor at the front. In addition to the five main buildings, there are also auxiliary buildings such as academies, auxiliary houses, stables, and back gardens. In addition to halls, patios and cloisters, there are more than 120 rooms, all of which have been built with bricks, stones and wooden structures.

◎ 尚书第第二幢天井。
The patio in the 2nd main building of Shangshu Mansion.

◎ 尚书第书院。
The academy of Shangshu Mansion.

◎ 尚书第南门。
The southern gate of Shangshu Mansion.

甬道贯通南北，连接南北大门。南门门额嵌有写着“尚书第”的巨幅石匾。北门是正门，门后有仪仗厅，是迎送宾客的地方。按传统建筑的一般规制，宅邸的南大门应该是正门，但尚书第的正门却在北边。李春烨为什么这么设计？人们猜测，或许因为京城在北，他这么做，是为了表示“面朝天阙”，感谢皇恩浩荡。

The corridor goes throughout the north and the south and connects the northern and the southern gates. There is a large stone plaque over the southern gate with the Chinese characters "Shangshudi" engraved on. The northern gate is the main entrance, and there is a ceremonial hall behind the door, where guests are met and seen off. According to the general standard of traditional architecture, the southern gate of the mansion is supposed to be the main entrance, but the main entrance of Shangshu Mansion is the northern gate. Local people speculated that it was because the imperial capital had been located in the North. Li Chunye had intended to express his gratitude to the Emperor by having his home gate facing the imperial capital.

◎ 尚书第北门。
The northern gate of Shangshu Mansion.

◎ 义路。
Yi (Righteousness) Road.

◎ 礼门。
Li (Ritual) Door.

回到北门进来，第二幢主体建筑是宅邸的中心建筑，门楼宽达6.4米，高6.8米，采用通天石柱和条石梁枋支撑墙体，石柱间砌斗砖墙。门楼雕刻嵌满精妙的人物雕像，以及浮雕牡丹，镂空锦鸡、卷草、团花等各种寓意吉祥的图案。门楼对面是影壁，影壁墙下有石雕须弥座垫起的长条形花台。门口两侧的甬道上又各设置了景门，两个门内匾额上分别写的是“曳履星辰”“依光日月”，门外匾额分别写的是“义路”“礼门”。这一幢应该是李春烨本人居住和接待宾客的地方。

The central building of the mansion is the second main building one can see after entering the northern gate. The gatehouse is 6.4 meters wide and 6.8 meters high. It was built with stone columns and sliced stone beams to support the walls, and there were brick walls built between the columns. The gatehouse was engraved with exquisite figures and statues as well as embossed peony, hollowed-out pheasant, rolled grass, flowers and other auspicious patterns. Opposite to the gatehouse is the screen wall and there is a long flower stand padded by carved stone Sumeru thrones. Two doors are set up respectively on the two sides of the corridor. On the two inside horizontal inscribed boards there are two phrases which indicate the high status of the owner. The two outside ones read 'Yi Road' and "Li Door". "Yi" means righteousness and "Li" means ritual in Chinese and these two characters are among the core of Chinese culture. This space should be where Li Chunye lived and hosted guests.

尚书第的每幢建筑从一进到三进，都设计有大小不一的天井，以使庞大的建筑里的每个厅堂和房间都享受到阳光和流通的空气。天井之下有排水暗沟，设计科学，即使大雨天也不积水，保证了住所的干爽。而每两幢建筑之间，又有封火墙相隔，廊门相通，进与进之间，又以砖墙或楹门相隔，确保整体建筑既统一又相对独立，更起到防火作用。

关于尚书第，泰宁人还有段谈资。

李春烨写信回家让家人买地造宅，原本要买比现在的尚书第大一倍的地，但是乡人刁难，有的把地卖给了李家建祠堂，有的卖给了他舅舅。儿子写信向他抱怨，他的回答是："官大大不过娘舅，争地不敢争祠堂。算了，就建一半吧。"

Every building in Shangshu Mansion is designed with a patio of different sizes so that every hall and room in the large group of buildings can get sun exposure and airflow. There are drainage ditches under the patios designed to ensure the dryness of the residence. Fireproof walls are built between every two buildings, while the corridors connect all the doors. There are also brick walls or doors to separate spaces, which ensure that the whole building is unified with considerable independence, while serving the purpose of fire prevention.

A story about Shangshu Mansion is passed among Tainning people.

Li Chunye wrote a letter to his family asking them to buy a piece of land to build a mansion on. Originally, he had wanted to buy a lot that was twice as big as the current Shangshu Mansion, but local residents purposely made it difficult. Some sold their lots to the Li's family for building an ancestral hall, and some sold theirs to Li's uncle. Li's son wrote to him to complain about the situation. Li replied, "No matter how senior I am as an official, I must respect my uncle. And I will never compete for the land intended for building an ancestral hall. Just make it half the size."

◎ 甬道。
A corridor.

尚书第是外人对它的称呼，李春烨给自己的宅邸起的名字其实是“五福堂”，因为只占到规划中一半的地，所以又被泰宁人叫作“半边福堂”。“五福”取意自《尚书·洪范》：“一曰寿，二曰富，三曰康宁，四曰攸好德，五曰考终命。”这是李春烨对自己和家族寄寓的美好愿望。

李春烨有5个儿子，李自枢最小，他8岁的时候，父亲李春烨就去世了，之后4个哥哥也相继去世，偌大的家宅家业就留给李自枢维持。他25岁中秀才，之后21年中，7次下福州乡试都没考上举人，最后在47岁时按惯例被授予了一个“州同”的小官，干了12年都没有升职，59岁心灰意冷地回家，守着空宅。晚年他写了一首诗，其中两句是“大笑谁能到百年，向平债绊弗恬然”。

富贵如云烟，人生如梦幻朝露，想想李春烨给自己宅邸起的名字“五福堂”，再念念李自枢的这首诗，站在气势雄伟的“五福堂”甬道上，设想李春烨晚年默默伫立其间的身影，不禁唏嘘。

◎ 门楼精美人物浮雕中间刻有“五福堂”三个字。

"Five-Fortune Mansion" can be seen on the carved stone plaque.

Li Chunye had named his home "Wufu Mansion" (Five-Fortune Mansion), the name Shangshu Mansion was given by others. Taining people also call it Half Fortune Hall since it only took half of the land of the original plan. The phrase "five fortunes" derived its meaning from the book *Shangshu•Hongfan,* "the first is longevity, the second is wealth, the third is peace, the fourth is kindness and conforming to nature, and the fifth is peaceful death". The good wishes of Li Chunye to himself and his family were embedded in this name.

Li Chunye had five sons, among which Li Zishu was the youngest. His father Li Chunye died when Li Zishu was 8 years old, and his four elder brothers died one after another, leaving Li Zishu alone to carry on the responsibility of taking care of the large family house and business. He succeeded in the imperial examination at the county level when he was 25 years old, but in the following 21 years, he failed the examinations at the provincial level in Fuzhou 7 times. Finally at the age of 47, he was appointed a junior officer according to custom without getting promoted for 12 years. At the age of 59, he went home with much disappointment and stayed in the empty house. In his late years he wrote a poem, which include two lines expressing his cynicism towards life rooted in such disappointment.

Wealth is like clouds and mist, and life is transient like a dream or the morning dew. One could not help but sighing, thinking about the name "Five-Fortune Mansion" that Li Chunye had given to his home and all his unfulfilled wishes.

从北门走出尚书第，门外就是现在的尚书街（古称“绣衣街”），也是旅游街。从历史的精神巡游中回到烟火人间，这时候最适宜做的事就是一饱口腹之欲。

在尚书街可以欣赏到擂茶姑娘做泰宁本地擂茶。她们手拿擂茶棍，在擂茶钵里行云流水般地转动搓揉。你则可叫上一份，加几小碟果蔬，端坐在老巷的古井边上，品味闲暇。

When you walk out of Shangshu Mansion through the northern gate, you will find a tourist street—Shangshu Street (known as "Xiuyi Street" in ancient times). You also return to the reality from the historical and spiritual tour. The best thing to do at this moment is to have a good meal to fill your stomach and satisfy your appetite.

On Shangshu Street, you can visit tea houses where you can view skilled tea-making girls make local tea of Taining called "pounded tea". They hold sticks in their hands for pounding tea leafs, maneuvering the sticks skillfully in the tea-beating container. All you do is to place an order, with a few dishes of fruits and snacks, and relax while sitting beside an ancient well in the historical lane.

◎ 尚书街。
Shangshu Street.

◎ 擂茶。
Making pounded tea.

◎ 游浆豆腐。
Fried tofu with soup.

◎ 暖菇包。
Mushroom dumplings.

◎ 乌株粳米糍。
Black rice cake.

◎ 碧玉卷。
Green stuffed rolls.

在尚书街，你还可以品尝到游浆豆腐、暖菇包、乌株粳米糍等，把泰宁的味道一股脑儿吃到肚子里，感受这实实在在的“福”，把尚书第深宅大院里的感慨通通抛到脑后去。

You can also try very special "swimming" tofu, mushroom dumplings, black rice cake and so on at Shangshu Street. After eating so much delicious food, you can feel the happiness from your stomach and leave all the deep thoughts and emotions about Shangshu Mansion behind.

◎ 明清园正门牌坊。

The stone arches at the entrance of Mingqing Palace.

收集记忆的明清园

明清园是“飞”来的景点，位于泰宁城区的五里亭，占地约220,000平方米，距尚书第约有4.5千米。

说它是“飞”来的，是因为这里的古建筑，包括园门口的石牌坊、园内的司马府第以及所有的古民居、古书屋、古书院、古厅堂等不同年代、不同风格、不同材质的古建筑都来自泰宁之外的不同地方，是整栋整栋被搬迁来的。除了古建筑外，藏宝阁里的古家具、古屏风、古牌坊、古戏楼以及诸多木雕、石雕、牌匾、楹联等，也都是从天南海北搜集到此的。它们“空降”泰宁，使泰宁的人文风情中有了与尚书第等本地建筑景观相辉映的外来气象。

明清园的主人公叫陈明青，福建尤溪人，是一位古民居和古木雕收藏大家。他从1983年开始从事收藏，从尤溪到广州，从广州到云南，再回到福建，最终落脚泰宁。几十年的积累，他从全国各地“抢救”下来几十组古民居和5000多件木雕珍品。

The Collector of Memories—Mingqing Palace

Mingqing Palace is an attraction that "flew in". It is located at Wuliting in Taining Town, covering an area of 220,000 square meters. It is approximately 4.5 kilometers away from Shangshu Mansion.

The reason why it is called a "flew in" attraction is that the ancient buildings here, including the stone arches at the entrance, Sima Mansion in the palace and all the ancient dwellings, book houses, academies, halls and other buildings of different times, styles and materials, all come from places outside of Taining and were moved here. Besides, the ancient furniture, screens, archways, opera houses and many wood carvings, stone carvings, plaques, couplets and so on were also collected from all over the country. Their landing in Taining brought more humanistic sense and different looks to the town. These buildings contrast with Shangshu Mansion and other local architectural landscape to bring out the glamour and radiance of Taining.

The owner of Mingqing Palace is Chen Mingqing, a great collector of ancient dwellings and wood carvings. He was born in Youxi of Fujian Province. He started his career as a collector in 1983 from Youxi, then went to Guangzhou, Yunnan, back to Fujian and finally settled in Taining. During several decades of accumulation, he saved dozens of groups of ancient dwellings and more than 5,000 pieces of wood carving treasures from all over the country.

◎ 明清园全景图。
A bird's eye view of Mingqing Palace.

司马府第是陈明青在江西某个旧料市场发现的，他装了几十车，浩浩荡荡地拉到昆明，然后花了3年时间拼接、修补，让这座始建于清朝中叶、占地1000多平方米的金碧辉煌的府第重现世间。

Chen Mingqing found Sima Mansion in a flea market in Jiangxi Province. He transported all the pieces to Kunming in dozens of trucks. He spent three years to splice and repair, getting this splendid mansion, which had been built in the middle of the Qing Dynasty that covered an area of over 1,000 square meters, back to life again.

◎ 司马府第大门。
The gate of Sima Mansion.

◎ 司马府第内景。
The interior of Sima Mansio

◎ 五百金身罗汉。

The screen of 500 golden-body Arhats.

五百金身罗汉原是清代道光年间云南一座商人家庙中的摆件。1997年陈明青花费几万元收购了这个摆件；过了两年，一个马来西亚华侨出价500万元购买，被他拒绝了。

The screen of 500 golden-body Arhats was originally a decorative piece in a family temple of a merchant in Yunnan during 1821—1850 in the Qing Dynasty. In 1997, Chen Mingqing spent tens of thousands of yuan to purchase it. Two years later, a Malaysian-Chinese offered 5 million yuan for it, but was refused by Chen.

◎ 聚贤堂照壁。

The screen wall of Juxian Hall.

“聚贤堂照壁”通体金丝楠木，以唐朝李世民的24位功臣为题材，展示了群贤聚集，忠信义勇，与君王同心同德的画面。据说这是清朝的一个军臣请人制作并准备上贡的，但在匠人作坊制作好、即将起运时，军臣因犯事被抄家，照壁因未进府第而幸免于难。君臣后人把这件家传遗物卖给了陈明青。

The screen wall of Juxian Hall was entirely made of nanmu wood which is rare. The storyline carved on the wall showcase 24 meritorious officials of the Emperor Li Shimin in the Tang Dynasty and their faithfulness to the country. It is said that the screen wall was prepared as a tribute by a military minister of the Qing Dynasty. Shortly before the scheduled shipping of the screen wall from the workshop to the military minister's house, however, the house was confiscated due to a criminal investigation. Luckily, the screen wall survived this event as it had not been brought to the official's house yet. The wall had become a family relic and was sold to Chen Qingming by the descendants of the military official.

2012年，明清园一期建设完成，这些藏品被悉数从昆明一车车搬到了这里。2014年，明清园被列为国家级文化产业重点支持项目，升级为国家AAAA级景区；2015年7月获得“中国木雕艺术博物院”称号。

现在明清园里有司马府第、聚贤堂、藏宝阁、耕读堂、碧桂斋、德馨书屋等10余个景观，与古典园林相互映衬，成了独具风韵的私家收藏文化大观园。

The first phase of Mingqing Palace was completed in 2012, and these collections were all transported here from Kunming. In 2014, Mingqing Palace was listed as a key project of the national cultural industry and became a 4A level tourist site. It was further awarded the title of "the Art Museum of Chinese Woodcarving" in July 2015.

There are more than 10 scenic sites in Mingqing Palace including Sima Mansion, Juxian Hall, Cangbao Tower, Gengdu Hall, Bigui Palace, Dexin Book House. They contrast with the ancient gardens, and have become a magnificent cultural pavilion of private collections with its unique charms.

◎ 古典园林。
A classical garden in Mingqing Palace.

岁月音符——中国大历史中的泰宁

“泰宁”这个名字是泰宁的第一位状元叶祖洽找皇帝赐的。

叶祖洽，泰宁城关叶家窠人，活了72岁。由于牵扯到“王安石变法”的新旧党争，官场一生是是非非。但他是泰宁第一个走向全国的人物，泰宁人为了纪念他，留下了一扇城门——昼锦门，而他为泰宁做的这件大好事也是被永远记住了。

1085年，北宋元丰八年，叶祖洽请他的好友，时任福建按察使的张汝贤奏请皇帝给老家泰宁——当时的归化县改名，给出的理由很有说服力。他是这么说的，“天下无水不朝东”，但山东曲阜孔府门前的泗河却向西流了三百里（约150千米），所以出了孔圣人；而归化县的金溪，从县城到梅口乡，向西流了三十里（约15千米），也是人杰地灵的宝地，当时的县名“归化”有“蛮夷归顺、王化”之意，“不正”。当时的皇帝宋哲宗准了，还将曲阜孔子阙里的府号“泰宁”赐为县名。这个县名一直沿用至今。这应该是泰宁在中国大历史中最成功的一次自我营销，毕竟这样一个偏远山区，可以这样被书写的机会并不多。

◎《泰宁赋》。
Taining Ode.

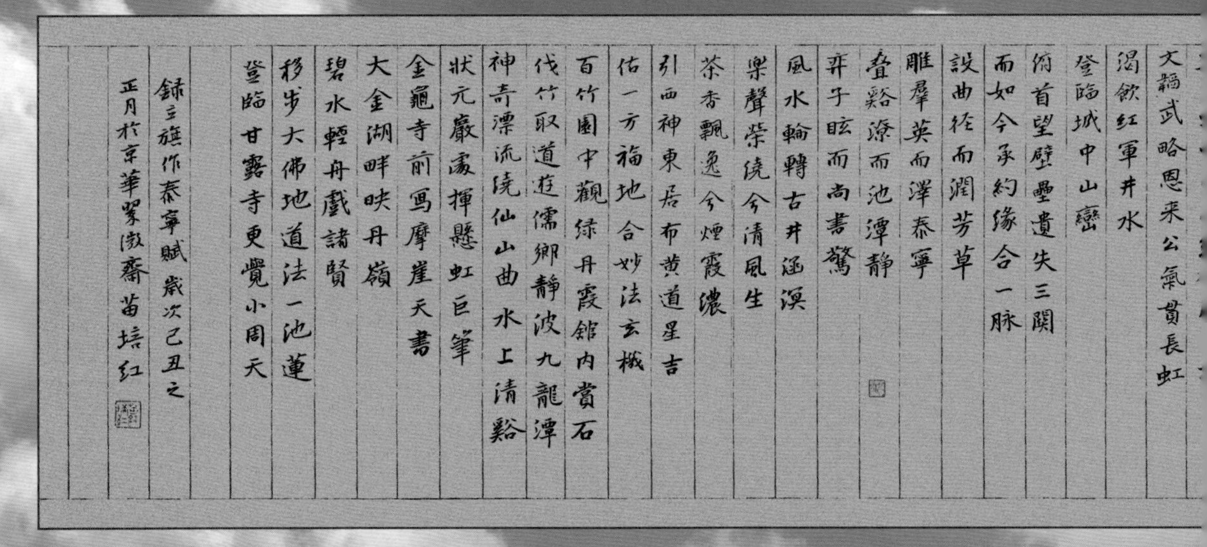

文韜武略恩来公氣貫長虹
渴飲紅軍井水
登臨城中山巒
俯首望壁壘遺失三關
而如今承約緣合一脉
設曲徑而潤芳草
雕群英而澤泰寧
叠谿潦而池潭静
弈子眩而尚書驚
風水輪轉古井涵溪
樂聲縈繞兮清風生
茶香飄逸兮煙霞濃
引西神東居布黄道星吉
佑一方福地合妙法玄機
百竹園中觀绿丹霞館内賞石
伐竹取道莅儒鄉静波九龍潭
神奇漂流繞仙山曲水上清谿
状元巖處揮懸虹巨筆
金龜寺前馬摩崖天書
大金湖畔映丹嶺
碧水輕舟戲諸賢
移步大佛地道法一池蓮
登臨甘露寺更覺小周天
録王旗作泰寧賦歲次己丑之
正月於京華翠微齋苗培红

泰宁在得到这个县名之前，很长一段时间被认为是蛮荒之地。因此，“泰宁”之前的县名为“归化”，即“蛮荒之地变成教化之邦”的意思。

泰宁得到“归化”这个名字，是在759年的唐朝肃宗时期，那时候还只是归化镇，958年才升为归化县。

759年以前，泰宁只是“场”，被叫作“金泉场”“金城场”，就是产金子的矿区。而在东汉之前，泰宁甚至没有一个专有地名。人们只是知道它在古时属于七闽，秦时属于闽中郡，西汉时属于闽越国。

这就是泰宁名字的沿革。回望而去，泰宁虽然只是藏在中国东南深山之中的山区小县，但它的历史无疑是悠久的。闪现在中国大历史各种典籍中的只言片语，以及遗留在泰宁大地的风物遗存，还是让我们找到了想象泰宁岁月长河中那些精彩故事的可能。

泰宁几年前在古城沿河修建公园，并创作了《千年赋》铜雕群，这是反映泰宁千年历史人文的19组大型青铜雕塑，记录了两千多年来泰宁历史中的重要人物和事件。对历史感兴趣的人，可以留宿在泰宁古城，晚饭后漫步在《千年赋》青铜雕塑间，来一次泰宁从古至今的穿越。

泰寧賦

千載靈秀動紫氣染丹霞
百代宏篇生瑞光照歸化
無諸校獵金鐃山
梅福煉丹棲真巖
鄒公開泰宗祠列章
頌漢唐戀古鎮之悠遠
憶兩宋覓明城之綿長
歷九夏出名宅尚書府第懷古
經三春入幽居世德堂前柱香
把酒壺盡尋覓處
卻醉於明城殘石
恍忽兮俊彩星馳
嗚呼哉回首經年
矚學東五魁坊
觀陽河兩狀元
一門四進士同賀
單卷九舉子共勉
李綱著易揚時倡道
朱子行吟邑留題四壁
遺琴封硯感梅雪太極
元宵春江唱十詠
日彩啟奏薦人雄
求言小引敬母賢
四世一品孝當先
閩越才俊傲天下

◎ 古城初雪。
The ancient town covered with thin snow.

The Note of Time: Taining in the Great History of China

"Taining" is a name given by an emperor. It was Ye Zuqia, Taining's first champion of the imperial examination that asked the emperor to give this name to his hometown.

Ye Zuqia was from Yejiake in Taining Town. He lived for 72 years, during which he had experienced a lot of ups and downs in his career because he was involved in the competition of the old and new parties in Anshi Wang's Reform. But he was the first Taining person who stepped onto the national stage. Taining people had left a gate called Zhoujin Gate to commemorate him. In the meantime, what he had done for the city is also remembered forever.

In 1085, the 8th year of Yuanfeng Period of the Northern Song Dynasty, Ye Zuqia asked his friend Zhang Ruxian, the judicial commissioner of Fujian, to request the emperor to rename his hometown—Guihua County. His reason was very convincing. He said, "Almost all water in the world flows toward the east." The Sihe River in front of Confucius Mansion in Qufu, Shandong Province however, flows to the west for 150 kilometers, and there born Confucius; while Jinxi Stream in Guihua County also flows to the west for 15 kilometers from the county center to Meikou Township, so it's supposed to be another treasure land propitious for giving birth to outstanding people. However, the name of the county was not auspicious since it means the naturalization of barbarians. As a result, the Emperor Zhezong granted the county the name "Taining", the same name of a building in Confucius's Mansion,

which shows the encouragement. This name has been in use ever since. This should be the most successful self-marketing of Taining in the history of China. After all, as such a remote and mountainous place, it has few opportunities to be written about like this.

The former name of the county was Guihua, meaning "the savage land was been civilized". So, the county had been considered a wild place for a long time before getting the new name, Taining.

Taining got its original name, Guihua, in 759 in the Tang Dynasty. At that time, Guihua was just a small town and was not promoted to a county until 958.

Before 759, Taining was only a "field" in administrative division. It was called "Golden-Spring Field" and "Golden-City Field", which means "a gold mining area". Before the Eastern Han Dynasty, Taining did not even have a proper name. People only know that it belonged to the Seven-Min in ancient times and to Minzhong Prefecture in the Qin Dynasty, and in the Western Han Dynasty, it was a part of Minyue Kingdom.

This is the history of how Taining got its current name. When looking back, we can find that although Taining is only a small mountainous county hidden in mountains, it undoubtedly has a long history. The flashing fragments in the various ancient books of the great history of China, as well as the remaining relics on the land of Taining allow us to find the possibilities of imagining the wonderful stories stored in the long history of Taining.

Several years ago, a park was built along the river in the ancient town of Taining, and the millennium bronze sculpture group was created. It is a large sculpture group of 19 bronze sculptures that reflect Taining's thousand-year history and culture, recording the important figures and events in its history. Those who are interested in history may stay in the ancient town for one night so that they can stroll around the millennium bronze sculptures and have a time-travelling of Taining from ancient times to the present.

金铙山下闽越王

秦汉之际的闽越国是福建有文字记载的历史的开始，也是泰宁有文可考的历史的开始。

闽越国王无诸，是我们所知的第一个闽王。无诸在今天的福州建了闽越王城，同时也在泰宁建了高平苑和乐野行宫。在古代，“宫”专指帝王的住所，“苑”是养禽兽植林木的帝王花园。也就是说，无诸当时经常跑到泰宁打猎，泰宁最早就是以“闽越王游猎之所”的身份，融入了整个福建的历史发展之中。

高平苑和乐野行宫在南宋时应该还留存着一些断垣残壁，南宋泰宁籍状元邹应龙的儿子、进士邹恕因此写了一首诗，“闽越遗宫蔓青草，萧萧衰柳满孤城。吟余独向荒台望，落日江山万古情。”经现代专家考证，乐野行宫的位置就在泰宁城区利达大厦处，高平苑的地点在泰宁水南食品公司处，但这些其实已经不那么重要了。

金铙山，在泰宁、建宁、宁化三县的交界处，属于武夷山脉的支脉大杉岭。主峰白石顶，海拔1857.7米，是福建境内的主要高峰之一。如果身在泰宁，要感怀一番“念天地之悠悠”，金铙山是首选。

清代著名诗人张际亮形容金铙山“纵横三百里，出没数千岫”。在白石顶眺望，壮丽的河山尽收眼底，如果能赶在东方渐露鱼肚时，看红日喷薄而出，由桃红、橘红顷刻间变成火红，千顷云浪一片金华，在脚下涌动，你会感觉仿佛来到神话中的仙界。

◎ 无诸校猎。

The King Wuzhu was hunting.

Jinnao Mountain and the King of Minyue Kingdom

Minyue Kingdom of the Qin and Han dynasties is the beginning of the history of Fujian Province as well as the beginning of Taining in the historical records.

Wuzhu, the king of Minyue Kingdom, is the first King of Min known to us. Wuzhu founded the royal city in today's Fuzhou, built Gaoping Yuan and Leye Palace in Taining. In ancient times, "Palace" specifically refers to the residence of an emperor. "Yuan" is a hunting ground for emperors. In other words, Wuzhu often came to Taining for hunting at that time. That is why Taining was integrated into the historical development as the role of "the hunting place of the King of Minyue" at the earlier period.

There should still be some ruins of Gaoping Yuan and Leyue Palace till the Southern Song Dynasty. Zou Shu, the son of Zou Yinglong, had written a poem about it, "The grass grows everywhere in Minyue Palace, and the rustling willows cover the city. I alone look to the direction of the desolate platform, only to see the sunset and mountains full of endless melancholy." According to the experts today, the original location of the palace is now the Lida Building in Taining Town and the location of Gaoping Yuan is now Taining Shuinan Food Company. But these are not so important.

Jinnao Mountain, or Gold-Cymbal Mountain, is located at the junction of Taining, Jianning and Ninghua. It is a part of the Danshan Peak of Wuyi Mountain. Its main peak is Baishi Peak (White-Stone Peak) with an altitude of 1,857.7 meters, which is one of the highest peaks in Fujian Province. If you are in Taining and would like to "experience the vastness of the world", Jinnao Mountain must be the first choice.

Zhang Jiliang, a famous poet of the Qing Dynasty, had described Jinnao Mountain "(a mountain) rolling for 300 *li* with countless caves". You can see the full view of mountains and rivers standing on the top of Baishi Peak. If you are fortunate to catch the early moment of the dawn when the red sun spurts out, you may feel as if you have come to a mythical fairyland. The sky's color changes from peach to orange, and suddenly turns red. All the clouds seem to put on gold coats and surge beneath your feet.

金铙山原本叫大戈山。“昔越王无诸常游猎于此”，校猎时，无诸很喜欢用铙帮忙。铙的声音很大，又很吓人，野兽最怕听到铙的声音。只要铙一响，野兽就会惊慌失措，甚至躲在一起，这时猎人就可以拉弓射箭，每次都会获猎满满。有一次，无诸的士兵不小心“遗金铙其上”。士兵们三五成群返回去找，只见远远的地方，铙在月光下金光闪闪。士兵们喜出望外，可是等他们走过去取时，铙又不见了，而是在另一个山头金光闪闪，再等他们走过去后又不见了。一连几个晚上都是这样。无诸说：“算了，铙是找不回来了，就让它留在这里吧。”

《泰宁县志》里记载，从此金铙山“至夜有光”，所以被称为“金铙山”。

另外，据很多史料记载，无诸死后葬在泰宁，清乾隆版《泰宁县志》记录了无诸墓的确切方位，还附了一首清朝苏州元和人施仁济到无诸墓前祭拜时写的诗：“遗窆传疑土一抔，霸图回首冶城秋。入关功在封何晚，横海军来战已收。落日荒原谁下马，丛芜古道自眠牛。耕童荛竖休轻触，风雨能添过客愁。”

因为无诸，泰宁的历史可以从秦汉时开始算起——距今已经有2000多年了。

◎ 闽越最高峰白石顶。
Baishi Peak, one of the highest peaks in Fujian Province.

◎ 远眺金铙山。

Jinnao Mountain seen from a distance.

The original name of Jinnao (Gold Cymbal) Mountain was Dage Mountain. Wuzhu, the king of Minyue, "always went hunting in Jinnao Mountain". He liked his soldiers to use the cymbals during hunting because the sound of the cymbals was very loud and scary. The wild animals were afraid of the sound. Once wild animals heard the sound, they would panic and even hide together. That's the time for hunters to draw the bows and shoot. By using this strategy, the soldiers could get a full hunt every time. On one occasion, the soldiers lost a cymbal in the mountain. When they returned in groups to look for the cymbal, they could see it shining golden light in the moonlight far away. The soldiers were overjoyed. But when they went to pick it up, the cymbal disappeared. It appeared and shined at another mountain top. After they arrived at the mountain top, it disappeared again. This happened for several nights. Wuzhu stamped with fury and said, "Well, forget it, you can not find it. Just let it stay here."

According to *Taining County Records*, since then, the mountain has been shining every night and that's why it is called Gold Cymbal Mountain.

Also, according to many historical documents, Wuzhu was buried in Taining after his death. *Taining County Records* (the Qing Emperor Qianlong version) recorded the exact location of Wuzhu's tomb and attached a poem written by Shi Renji, a poet of the Qing Dynasty from Yuanhe in Suzhou when he went to worship in front of the tomb.

Because of Wuzhu, the history of Taining can be dated back to the Qin and Han dynasties—it has been more than 2,000 years.

宝盖岩下，泯没成尘的开泰公邹勇夫

唐末五代十国时期是继汉末三国时期之后的又一乱世时代。这期间王审知兄弟起兵入闽。后王审知受封为闽王，派手下大将左仆射邹勇夫率军防守闽西北，驻军所在地就是归化（今泰宁县城）。

邹勇夫到泰宁时，看到的是“榛芜亘野，烟火仅百家”。他采取“先靖后治”的对策，抚集四方流亡，帮助他们修理房舍，鼓励他们开垦荒地，引进中原先进生产技术，并兴学校、敦教化，于是“民襁褓而至，始遂生息”。自此以后，泰宁文教、经济、人口都才算有了些样子，泰宁成为闽西北一大县邑。邹勇夫因此被称为“开泰公”。

邹勇夫922年逝世，葬于泰宁县邹家寨。邹家寨就在宝盖岩，在泰宁朱口镇的寨色村境内，距离县城15千米左右。宝盖岩是泰宁世界地质公园四个园区之一的石辋东南方的一个隘口所在。驴友去石辋大峡谷玩时，很多会选择从宝盖岩下进入。

◎ 开泰公邹勇夫。
A statue of Zou Yongfu.

Baogai Rock, Carrying the Achievements of Mr Kaitai—Zou Yongfu

The Five-Dynasties (907—960) and Ten-Kingdoms (902—979) period at the end of the Tang Dynasty was another chaotic period. During this period, Wang Shenzhi and his brother sent an army to Fujian Province ("Min" for short). Later, Wang was made the king of Min. He then sent his general Zou Yongfu to lead the army to defend the northwest of Min. It was Guihua (today's Taining) where the garrison was stationed.

When Zou Yongfu first arrived at Taining, what he saw was "the weeds grew all over the mountains and fields, and there were only hundreds of families". So, he adopted the strategy to "pacify first and then govern". He gathered the people in exile and helped them repair their houses, encouraged them to reclaim the wastelands, introduced the advanced production technology of Central China to Taining, and set up schools, putting focuses on education. After that, the rate of population increased and the economy and culture of Taining had made rapid progress. Taining had grown to a large county in northwestern Min. Zou Yongfu was therefore called "Mr Kaitai", meaning that it was Zou Yongfu who brought Taining a new start.

Zou died in 922 and was buried in Zou's Village in Taining County. Zou's Village is located at Baogai Rock. It's about 15 kilometers away from the county town. Baogai Rock is the southeastern mountain pass of Shiwang, one of the four parts of Taining Global Geopark. Many hikers would choose Baogai Rock as the entrance to Shiwang Grand Canyon.

◎ 宝盖岩。
Baogai Rock.

据《泰宁县志》记载，早在公元923年五代十国时期，就有一位叫定慧的四川僧人云游至宝盖岩，在此结草为庵。宝盖岩寺于南宋时期始建，算来有1000余岁“高龄”了。宝盖岩不高，宝盖岩寺位于半山腰，上下均为陡峭的岩壁。沿右侧石径而上，有前后两座城堡似的山门扼守要冲，前面一座为清乾隆五十七年（1792）建，匾额上题刻“烟霞洞”三字。旧时每遇兵匪劫乱，山门一闭，就可摒蟊贼于山外。

According to Taining County Records, as early as 923 in the Five-Dynasties and Ten-Kingdoms period, there was a Sichuan monk, Dinghui, who traveled to Baogai Rock and built a nunnery there. Baogai Rock Temple was built in the Southern Song Dynasty with a history of more than 1,000 years. It is not tall, and it is located halfway up the mountain with steep rock cliffs along the way. Walking along the right side of the stone path, you can see two castle-like gates guarding the front and the back sides of the temple. The front one was built in 1792, with a board with "Yanxia Cave" on it. In the old days, whenever the looting or pillage occurred, the residents can easily protect themselves by closing the gates.

◎ 宝盖岩宝殿。
Baogai Rock Palace.

◎ 舍利岩。
Rock of Buddhist Relics.

岩寺的右侧还有两个较大的岩洞，一个叫“舍利岩”，另一个叫“谷仓岩”。舍利岩内建有清康熙以来岩寺历代高僧的舍利宝塔十余座。在谷仓岩路下的悬崖峭壁间，又有一个小岩洞，洞中就是邹勇夫的崖葬遗址。听说邹氏子孙们每年清明节会携香烛供果入洞祭拜。

除此之外，邹勇夫在泰宁似乎并没有留下其他痕迹。即便是他墓葬所在的宝盖岩寺，也没有因为他而声名远扬。倒是他的后代、泰宁宋朝时出的两个状元之一的邹应龙，在泰宁享受了更多的荣耀。卒后，以赙典皇室仪式敕葬“南谷”山岗，俗名“状元冢”。邹应龙小时候读书的山洞被命名为“状元岩”，现被开发为泰宁著名的旅游景点，游人不绝；邹应龙的后代更是在其出生地泰宁县城水南街南谷巷筹资兴建了气势宏伟的邹氏祠堂——南谷堂。祠堂大殿供奉着三位邹氏祖先，中间的是邹应龙，邹应龙右边的便是邹勇夫。

这可能是因为邹应龙算是泰宁历史上出过的最大的官吧，官拜端明殿学士、签书枢密院事、权参知政事，算是副宰相，而且他的子孙人数庞大，有数百位之众，他因此还被封了神——广佑圣王，信众遍布闽西、闽南、广东、台湾及东南亚地区。

不管怎么说，“开泰公”邹勇夫来守泰宁，城邑初备，文教始开，这才有了后来的故事。

邹应龙到过宝盖岩，肯定凭吊过他的祖先邹勇夫，不过没有留下什么文字，只留下一首《游宝盖岩》：“夙有兹岩约，今晨得践盟。路从支涧入，人在半空行。六月如霜候，四时长雨声。愿求容膝地，着我过浮生。”

There are two large caves on the right side of the temple, one is called "Rock of Buddhist Relics" and the other is "Barn Rock". More than ten stupas of eminent monks were built in the rock temple since the Kangxi Period of the Qing Dynasty. Between the cliffs under Barn Rock is a small cave, where is the cliff burial site of Zou Yongfu. It is said that the descendants of Zou would come and worship him at every Tomb-Sweeping Festival.

It seems that Zou Yongfu had not left any other traces in Taining. Even the temple where he is buried was not famous for him. However, his offspring Zou Yinglong had enjoyed more glory in Taining. After his death, Zou Yinglong was buried with a royal ceremony in South Valley Hill, which was later called "Zhuangyuan (Champion) Grave". The cave where he used as a learning place when he was a child was called "Zhuangyuan Rock" and now it has been developed as a famous scenic spot of Taining. The descendants of Zou Yinglong also have raised funds and built an impressively large ancestral hall for Zou Yinglong at his birthplace. Three ancestors of Zou's family were enshrined in the hall. It is Zou Yinglong's statue that sits in the middle, with Zou Yongfu's on the right.

Zou Yinglong was apotheosized as God—Guangyou King and his believers are around the east and west of Fujian, Guangdong, Taiwan and Southeast Asia. And it was Mr Kaitai—Zou Yongfu, who first came to Taining, defended Taining, established the city, promoted the development of culture and education of Taining and opened a new chapter in the history for Taining.

◎ 邹勇夫崖葬地。

The cliff burial site of Zou Yongfu.

◎ 朱熹铜像和《四季壁诗》。
The statue of Zhu Xi and his poem.

小均坳，朱熹的背影

泰宁县博物馆内珍藏着南宋理学家朱熹的四块手迹碑刻，这是他晚年创作的一首诗，内容是吟咏春、夏、秋、冬四景：

“晓起坐书斋，落花堆满径，只此是文章，挥毫有余兴。

古木被高阴，昼坐不知暑，会得古人心，开襟静无语。

蟋蟀鸣床头，夜眠不成寐，起阅案前书，西风拂庭桂。

瑞雪飞琼瑶，梅花静相倚，独占三春魁，深涵太极理。”

诗文没有落款，因为写这首诗的时候，朱熹正在泰宁避难，不能太张扬。庆元三至四年间（1197—1198），朱熹是为躲避“党禁”事件的迫害来泰宁隐居的。但诗中依然可感受到作者心境的恬淡如常，要知道此时的朱熹，不仅背负危险，更已经是一个病痛缠身、年近70的垂暮老人，这时距离1200年他去世也不过两三年的时间。

据《泰宁县志》记载，南宋庆元年间，朱熹被朝中权贵视为“伪学逆党”而遭受迫害，四处躲避，期间在小均坳住了一段时间，在当时泰宁知县赵时馆的暗中保护下，得以继续建筑书舍授徒，并写下了这《四季壁诗》。

清朝时，有人整理了《杉阳名胜小记》，列了泰宁二十六景，小均坳也在其中，名为“考亭琴涧”。书里把这个地方描绘得很美，“密竹古藤，篱落数椽，自为烟火”，还说朱熹走后，“遗有像一幅，砚一方，琴材一片。后乡人置材跨圳以渡，小凫凫过音生，山水皆响”，意思是说，乡亲们后来把朱熹本来用来做琴的木料铺在小山涧上当桥，小鸟飞过时，桥会发出声音，连同这里的山水也发出了声响。

朱熹离开泰宁后，书舍渐渐荒废，《四季壁诗》落为农家收藏。清乾隆年间，泰宁教谕李开将《四季壁诗》藏于文庙。后因年久墙废，碑文断裂，于20世纪80年代被泰宁县博物馆珍藏。

Xiaojun'ao, a View of Zhu Xi's Back

The Taining County Museum houses four inscriptions of Zhu Xi, a Neo-Confucianism scholar in the Southern Song Dynasty. It is a poem he wrote in his later years, chanting the scenery of the four seasons:

(I) got up early in the morning, sat in the study reading a book and noticed the small road outside was full of falling flowers. (I) burst into high interest and began to write articles. (Spring)

The ancient trees were forested with shades, where I cannot feel the hotness when I sit under. I opened clothes, sat quietly, read a book written by the ancients, and tried to understand what they had experienced and wanted to tell readers. (Summer)

In the autumn nights, crickets kept chirping and I couldn't fall asleep. So, I'd rather sit in front of my desk and read my book. The west wind blew in the yard, bringing the fragrance of the Osmanthus to me. (Autumn)

In the white world of winter, the plum blossomed charmingly, reporting the coming of the spring. See, everything in the world, four seasons, Yin and Yang movement, the universe of Tai Chi, follows the Neo-Confucianism. (Winter)

◎ 朱熹四季诗手迹。
The handwriting of Zhu Xi.

Zhu Xi did not sign his name at the end of the poem because when he wrote it, he was taking refuge in Taining, and he did not want to give himself away. During 1197—1198, Zhu Xi came to Taining to live in seclusion to escape the persecution caused by the "Ban of Parties". However, we still feel the calmness of the poet in the poem. It is necessary to know that when he wrote this poem, the poet was not only in danger, but also an old man who was nearly 70 years old and suffered from illness. He then died two years later in 1200.

According to *Taining County Records*, during the Qingyuan Period of the Southern Song Dynasty, Zhu suffered persecution because his theory of Neo-Confucianism was regarded as the pseudo-theory by the influential officers. He was forced to venture everywhere to find a shelter. He lived at Xiaojun'ao for a period during that time under the protection of Zhao Shiguan, the magistrate of Taining. He founded a school and continued giving lectures to students and wrote the above-mentioned poem there.

In the Qing Dynasty, someone compiled a book titled *A Note of the Attractions in Shanyang* and listed the 26 attractions in Taining, including Xiaojun'ao. This place was described as an amazing place: "(there are) dense bamboos, ancient vine, citrons, and dwellings." It is also said that Zhu left there a painting, an inkstone, and a piece of wood that can be used for making a Chinese music instrument. About the wood, the book also records another story: the villagers later used the wood to build a bridge on the stream. Afterwards, every time a bird flew over, the bridge, together with all the mountains and rivers around, would make a beautiful sound.

After Zhu Xi left Taining, the school fell into disuse and the inscriptions of Zhu Xi's *Four-Season Poem* became a farmer's collection. During the Qianlong Period of the Qing Dynasty, Li Kai hid them in a Confucian Temple. But the inscriptions were broken because of the ruin of walls. Then they were collected by Taining County Museum in the 1980s.

石鞨南石寨，历史的尘埃

石辋是泰宁世界地质公园的四个地质景观园区之一，在泰宁东北十余千米的朱口镇石辋村，具有和周边地区截然不同的丹霞网状谷地和山地，其中发育有雄伟壮观的丹霞峰柱地貌，构成了枪山、旗山、牌山、孝子峰等壮丽景观，因地势险要，一直是历朝历代战乱时泰宁乡邻避乱的首选地点。

明代文人池显方写过，“石辋者，石周遭如城，而自辟五门，内村原十里，类桃源者也……”，“自辟五门”意思是说天然地势形成了五个与外界隔绝的隘口，峭壁雄关，易守难攻，里面有村落，仿佛世外桃源。

石辋大峡谷还是未经旅游开发的处女地。去游石辋，可以选择从正南方隘口的宝盖岩下进入，也可以经由泰宁长兴村，穿越一片稻田，沿着山涧进入。

◎ 枪山。
The Gun Peak.

◎ 旗山。
The Flag Peak.

◎ 孝子峰。
The Dutiful Son Peak.

Nanshi Fortress in Shiwang—a Piece of Sad History

Shiwang is one of the four geological landscape parks among Taining Global Geopark. It is located in Shiwang Village of Zhukou Town, which is more than ten kilometers northeast of Taining. It has Danxia netlike valleys and mountains that are completely different from the surrounding areas. There are majestic Danxia peaks and landforms, which constitute the magnificent landscapes of the Gun Peak, Flag Peak, Cards Peak and the Dutiful Son Peak. Due to the difficult terrain, it has always been the preferred location for Taining people to avoid rebellions through the dynasties.

Chi Xianfang, a literatus of the Ming Dynasty, once wrote about Shiwang. According to him, the natural terrain has formed five mountain passes that were isolated from the outside world. The majestic cliffs make Shiwang easy to defend but difficult to attack, resulting in a paradise-like village within.

Shiwang Grand Canyon is still a virgin land without any tourism development. When you come and visit Shiwang, you can choose to enter from the southern entrance of Baogai Rock or go through a rice field and then go along the mountain creek in Changxing Village of Taining.

泰宁的历史来到明清之际，世人的目光聚焦之处就是石辋。这又可以从金庸格外推崇的那位明末大将袁崇焕讲起。

1622年，袁崇焕在福建邵武当知县，泰宁当时是邵武管辖的一个下属县。而泰宁籍进士、正四品高官江日彩，在这一年正月给皇帝上了一个奏折《议兵将疏》，建议对袁崇焕“破格议用”。

江日彩和同时代的泰宁籍的另一个大人物李春烨是同学，两人的老师就是江日彩的父亲，后来两人还成为亲家，江家女儿嫁给了李家儿子。他们同在1606年中举，第二年又一起进京会试，江日彩中进士，李春烨却到9年后才考上。后来李春烨因附和魏忠贤青云直上，江日彩在做官的最后几年却因为看不惯魏忠贤的飞扬跋扈要上表辞官回家。

江日彩应该是在平时回泰宁老家探亲、路过邵武府城时与袁崇焕认识，并对他的军事才能和见地有深刻印象。

1622年，江日彩正月上奏，二月袁崇焕就真的被破格提拔去兵部，很快又到了辽东前线抵抗步步紧逼的清军。在他的指挥整合下，腐败无能的明朝辽东部队在短短时间内就成了让敌人胆寒的劲旅。1626年，宁远城下一万孤军抵挡努尔哈赤带领的十几万大军，努尔哈赤一生驰骋，就只遭遇这一次败仗，不久就死了。但很快，1630年，袁崇焕被皇太极以反间计迫害，被崇祯皇帝以通敌的罪名凌迟处死。

袁崇焕死后14年，即1644年，清军入关，攻陷北京，明亡。1645年发生了一次著名的历史屠戮事件，即“嘉定三屠”。这个事件的主导者是明降将李成栋。1646年，这个李成栋继续率军攻入福建，先攻邵武，再攻泰宁。泰宁城区和周边地区的两万多民众相继避难于石辋，组织他们闭关自守、抵抗清军的是江日彩的两个儿子——江豫和江复。

史书上说，江豫“耿介豪迈”，江复“谦谦君子”，两个人都是诸生，也就是在州府县或者国子监读书的生员。江复做过一首诗《闻国变》，“英雄无可奈何处，白尽头颅只自憎。”

这时候，江日彩已经去世21年，时隔他举荐袁崇焕已经24年了。江日彩举荐袁崇焕抗清和他两个儿子抵抗清军，分别谱写了两段同样惊心动魄且让人黯然神伤的篇章。在泰宁，江氏家族因此极受尊重，被誉为“满门忠烈”。

◎ 御史荐才。
Jiang Ricai recommended Yuan Chonghuan to the emperor.

Shiwang has stood out in the history of Taining during the Ming and Qing dynasties. Yuan Chonghuan, the famous general of the late Ming Dynasty who was especially admired by the famous writer Jin Yong, had his stories spread in Shiwang.

In 1622, Yuan was the magistrate of Shaowu County in Fujian Province with Taining under its jurisdiction. In the first lunar month of that year, Jiang Ricai, the senior officer of the fourth level from Taining, submitted a memorial to the Emperor to suggest an exceptional promotion of Yuan.

Jiang Ricai and Li Chunye, another important figure of Taining, were classmates and shared the same birthplace. The teacher of the two was Jiang Ricai's father. Later, the two also became relatives because the daughter of Jiang married to the son of Li. They both succeeded in the provincial imperial examination in 1606 and went to the capital city to take the following exams. Jiang became Jinshi but Li failed. Until nine years later, Li succeeded again. Then, Li was promoted extremely fast because of the relationship between him and Wei Zhongxian, while Jiang requested retirement in the late years of his career because he cannot stand the domineering Wei.

In 1622, after Jiang Ricai submitted the memorial in January, Yuan Chonghuan was promoted to the military department in February, and soon was sent to the front line in Liaodong to resist the troops of the Qing Dynasty. Under his command and integration, the corrupt and incompetent troops of the Ming Dynasty turned into a fearsome force in a short period. In 1626, under the command of Yuan, Ningyuan City's 10,000 soldiers resisted the hundreds of thousands of troops led by Nurhachi, which was the only defeat Nurhachi encountered in his lifetime. Nurhachi died a short time after the defeat. But soon in 1630, Yuan Chonghuan was persecuted by Nurhachi's son Huang Taiji and was executed by the Emperor Chongzhen on the charge of colluding with the enemy.

In 1644, 14 years after Yuan's death, the troops of the Qing Dynasty captured Beijing. The Ming Dynasty went to its end. In 1645, the known "Jiading Slaughter" happened. The leader of the slaughter was Li Chengdong, the surrendered general of the Ming Dynasty. In 1646, Li Chengdong led his army to enter Fujian, attacking Shaowu first and then Taining. 20 thousand people in Taining and surrounding areas had taken refuge in Shiwang. It was two sons of Jiang Ricai—Jiang Yu and Jiang Fu that organized these people to defend themselves and resist the troops of the Qing Dynasty.

According to the historical records, Jiang Yu was upright and had a heroic spirit while Jiang Fu was a modest cautious gentleman. Jiang Fu once wrote a poem titled *Hearing the Dynasty's Change* : "Even the heroes feel hopeless, (they) can only let the hair turn white and hate their powerless."

When the two sons resisted the army of the Qing Dynasty, their father Jiang Ricai had been dead for 21 years. And 24 years had passed after Jiang Ricai recommended Yuan Chonghuan to resist the army of the Qing Dynasty. Jiang's family was highly respected and was known as "valiant warriors" in Taining.

李成栋进攻石辋，很快门户被击破，泰宁人转而据守石辋最险峻的南石寨。

南石寨，因“岩崖峻险，石皆南向”而得名，其上有岩穴旷地，可容纳数万人。入寨的道路是沿崖壁狭窄的山脊凿出的石阶，隘门在半山腰，极为险要。在这里，由乡民临时凑成的抵抗队伍和李成栋的正规军展开了惨烈的搏斗。寨破，江豫战死，江复慷慨就义，接着就是一场泰宁历史上空前绝后的大屠杀。包括妇孺老弱，被杀及跳崖死者有万人以上，据说占了当时泰宁全县人口的四分之一到三分之一。

漫地血光之后，生产力遭受了毁灭性的破坏，但更连根拔起的毁灭其实是在人文上。当百年之后，人口恢复，泰宁的文化和人才在清代却再也没有多少可称道之处。

历史总是太过宏大，发生在泰宁石辋南石寨的这一场大屠杀，除了泰宁的县志，似乎没有被其他的史书提到过。但是，如果你来到泰宁，来到石辋大峡谷，来到南石寨，站在南石寨所在的丹霞崖顶之上，看远处群峰叠翠，近处树木青葱，一切复归宁静，早已找不到任何惨烈战斗的痕迹，可你还是应该听听这些尘封的故事。

这是历史，也是一片土地背后令人感慨的命运。风景壮丽，群山无言。

◎ 石辋南石寨遗址。
The relics of Nanshi Fortress in Shiwang.

Li Chengdong's army soon broke the portal of Shiwang. Taining people had to turn to the most precipitous place—Nanshi Fortress and continued to resist.

The fortress was called Nanshi (South-Stone) Fortress for "all the stones there faced south and were precipitous". There were plain grounds on the top of the rock, which could accommodate tens of thousands of people. The access to the fortress was stone steps that were cut along the narrow ridge of the cliff. The safety gate was halfway up the mountain and was extremely dangerous. Here, the resistance team temporarily formed by the villagers and the army led by Li Chengdong launched a bloody fight. The resistance team failed, Jiang Yu fought to death, and Jiang Fu went to his death unflinchingly. Then, there occurred an unprecedented massacre in the history of Taining. More than ten thousand people were killed or jumped off the cliff. The death accounted for a quarter to one-third of the population in Taining.

After the slaughter, productivity suffered devastating destruction, but the humanities suffered more. Even though the population returned to a regular range after a century, the culture and talents of Taining were not worth mentioning in the Qing Dynasty.

History has included too much. The slaughter happened at South-Stone Fortress seemed never be recorded by other history books except *Taining County Records*. When you stand on the top of the Danxia cliff where the fortress was located, looking at the numerous distant peaks, green trees nearby, you will not find any traces of the fierce battles. Everything has come back to peace. But you must know the history.

◎ 石辋大峡谷。
Shiwang Grand Canyon.

◎ 红军街。

The Red Army Street.

红军街往事，“山路弯弯，红星闪闪”

清代至民国初期，泰宁在全国一直没有什么存在感，它沉寂了。等它重新获得世人的关注，已经是20世纪30年代。

当时泰宁流传着一首歌谣：“油菜开花七寸心，剪掉辫子当红军。保护红军万万岁，牺牲性命也甘心。”泰宁城区内的岭上街，现在称为“红军街”，是炉峰山下一条明清风格的古巷。走进红军街，你的心绪会回到八十多年前。

泰宁这座小城，在经历了长时间的平静之后，在20世纪30年代进入了一个激荡的历程。这里从1931年起成为中央苏区县，并一度成为闽赣省物资供应和经济文化建设的中心。

The Past of the Red Army Street

From the Qing Dynasty to the early Republic of China, Taining had been silent. When it regained the attention of the world, it was already in the 1930s.

There was a song spread in Taining singing that local people were anxious to join the Red Army at that time. People would like to cut their braids to join the Red Army and sacrifice themselves to protect the Red Army. The street now called Red Army Street was originally Lingshang Street in the town area of Taining. It is an ancient lane of the Ming and Qing styles. Walking on the Red Army Street, your mind may go back to more than 80 years ago.

This small town entered another turbulent historical period in the 1930s after a long period of silence. Taining had been a base county for the Red Army's central government since 1931 and became the center of material supply as well as an economic and cultural construction center in Fujian and Jiangxi provinces.

泰宁在地理上，是中央苏区东北方向的重要屏障、福建苏区与闽赣苏区的主要通道，因此成为闽赣边区战略要地，是军事争夺上的重要目标。从第二次反“围剿”到第五次反“围剿”，红军曾分别于1931年6月、1932年10月、1933年7月三次入泰宁，期间有近3000名泰宁人参加了红军。

岭上街12号陈家大院，“朱德、周恩来同志故居”的牌匾高悬在门额上。1933年8月，朱德、周恩来率领的红军总部及红一方面军总部从江西向东转移，经建宁抵达泰宁，便在这陈家大院里指挥红军“赤化千里，筹款百万”。

陈家大院是一栋建造于清朝初期的三进民房，三进厅堂分别设有二十余间厢房，整体建筑为砖木结构，四周筑有防火墙。当时，朱德住后厅左侧，周恩来住前厅左侧，杨尚昆卧室紧挨边上。院后山坡上还挖了一个简易的防空洞，以备战时之需。

Geographically, Taining is an important barrier in the northeast and the main channel between Fujian and Jiangxi. Therefore, it had become a strategic location and an important target for military competition. From the 2nd to the 5th Anti-encirclement Campaign, the Red Army entered Taining three times in June 1931, October 1932, and July 1933, during which nearly 3,000 Taining people joined the Red Army.

Chen's House is the former residence of some important leaders of the Red Army. A plaque hangs high on the top of the door header and writes, "The former residence of Zhu De and Zhou Enlai". In August 1933, Zhu De and Zhou Enlai led the headquarters of the Red Army to move eastward from Jiangxi Province. After arriving in Taining, they settled down in Chen's House which was built at the beginning of the Qing Dynasty and there were more than 20 rooms.

◎ 朱德、周恩来同志故居。
The former residence of Zhu De and Zhou Enlai.

红军街老墙上的那些口号见证着那段岁月：“打倒帝国主义!”“只有武装动员起来，实行土地革命，用武装拥护苏联”……最引人注目的是杨尚昆起草的巨幅文告《告刘和鼎部下士兵及下级官长书》，高2.6米，宽4.2米，全文665字，为苏区保存最完好、面积最大的文字遗迹。

The slogans on the old walls of the Red Army Street witnessed the past years: "Down with imperialism!" "Only through armed mobilization and implementation of the agrarian revolution can we support the Soviet Union" ... The most striking one is the huge statement drafted by Yang Shangkun, *A letter to the soldiers and lower-level officers of Liu Heding*. It was 2.6 meters high and 4.2 meters wide with 665 words. It is the best-preserved and largest text relic in the Red Army's bases.

© 红军街标语。

The slogans on the old walls of the Red Army Street.

登临相邻的炉峰山，还可以看到红军当年修筑的战壕以及烈士纪念亭。不仅如此，还能看到东方军司令部旧址，在泰宁县城水南金富街罗汉寺的观音阁内。罗汉寺建于五代年间，距今已有1000余年历史。此外，还有红军医院、苏区银行、红色广场等众多革命遗址遗迹，这些都会让人感受到这里曾经的滚烫岁月。

Climbing on the adjacent Lufeng Mountain, you can also see the trenches built by the Red Army and the Martyrs Memorial Pavilion. Besides, you can see the former site of the Eastern Army Command which is located at Guanyin Pavilion of Luohan Temple on Jinfu Street. Luohan Temple was built in the Five-Dynasties period with a history of more than 1,000 years. Also, there are many relics and former sites of the revolution such as the Red Army Hospital, the Bank of the Su-District, the Red Square, which will make people feel the burning passions of those years.

◎ 告刘和鼎部下士兵及下级官长书。
The letter to the soldiers and lower-level officers of Liu Heding.

◎ 东方军司令部遗址。
The former site of the Eastern Army Command.

洞里乾坤——灵山秀水间的“一千零一夜”

泰宁被称为“丹霞峡谷大观园”“丹霞岩穴博物馆”，它们是泰宁独特的地理标签，也是这片土地独特文化的承载。

泰宁的青年期丹霞地貌的特点之一就是崖壁上洞穴众多，孕育了泰宁独有的岩穴人文景观，这是这里的地理与这里的人在漫长的生存探索中相适应的结果。千百年来，泰宁重要历史人物的出场及影响泰宁历史进程的很多方面，都与岩穴有关。泰宁岩穴里，藏着泰宁的历史，藏着属于泰宁的“一千零一夜”故事。

泰宁的学子都会选择一个自己认为是福地的岩穴，甘受寂寞，潜心读书，借助山川灵气启迪文思，从而打开科举之路。叶祖洽中状元前，在叶家岩苦读；邹应龙则是在现在被称为“状元岩”的地方苦读——他背粮而凿的斗米阶，引领他一直到达科举的顶峰；明朝做官做到太仆寺少卿的江日彩在黄石寨的人干岩；兵部尚书李春烨则在李家岩。这种文化一直延续到清末废除了科举制度，泰宁仍有不少村落把丹霞洞穴作为书馆经堂，如际溪的丰岩、南会的云岩、丰岩的赤溪。新中国成立后，丰岩的赤溪小学有一段时间仍在岩穴中上课。

岩穴还收留落魄文人、出世隐士，让他们“托性于山林，寄情于物外”，在冷峻的岩穴之中洞悉人生一场空梦。宋抗金名将李纲被罢相，贬到海南岛，回来后就在曾在泰宁丰岩问道于禅师；明末的邱嘉彩，在明亡之际，跟着明将傅冠举兵勤王。后来兵败，邱嘉彩就举家隐居岩穴，“尺地可安生，带妻孥偕隐；高天堪问，与日月以争光”，20多年不入城。后来人们赞其“国之肖臣，母之肖子”，所以他住过的岩穴就被命名为“肖岩”。

◎ 肖岩。
Filial Piety Rock.

The Spiritual Universe in Taining: "One Thousand and One Nights"

Taining is known as "the Grand View Garden of Danxia Canyon" and "the Museum of Danxia Caves". They are the unique geographical labels of Taining, bearing the unique culture of this land.

One of the characteristics of Taining's youth Danxia landform is that there are many caves on the cliff walls, which have nurtured the unique humanistic cave landscape in Taining. This resulted from the adaptation between the geology and people's long-term exploration for survival here. For hundreds and thousands of years, the appearance of important historical figures and many aspects affecting the historical development of Taining have been related to local caves. The history of Taining and the story of "One Thousand and One Nights" are hidden in these caves.

In the past, all the students of Taining would choose a cave that they thought was a place of fortune to study alone and diligently to be enlightened by the spirit of mountains and begin their journey to imperial examinations. Ye Zuqia studied hard at Ye's Rock before he succeeded in the examination; Zou Yinglong beavered away at today's Zhuangyuan Rock. The Doumi stone steps that he chiseled for carrying food supplies had led him to the peak in the imperial examination. Rengan Rock was the study place of Jiang Ricai, who became an official in Taipushi equivalent to the Ministry of Transportation today in the Ming Dynasty; Li's Rock was the place where Li Chunye studied and became the minister of the military department. This culture continued until the end of the Qing Dynasty when the imperial examination system was abolished. After that, however, there were still many villages near Danxia caves and the caves were used as libraries or sutra halls, such as Feng Rock in Jixi, Cloud Rock in Nanhui, Feng Rock in Chixi. Even after the founding of the People's Republic of China, Chixi Primary School in Feng Rock still had lessons in the caves for some time.

The caves also sheltered sorehead literati and recluses, allowing them to "shift their attention and feelings to nature" and to understand in the cold caves that life was nothing but a visional dream. Li Gang, a famous general of the Song Dynasty, had once learnt from a Buddhist monk at Feng Rock in Taining on his way to Hainan Island after being dismissed. At the end of the Ming Dynasty, Qiu Jiacai followed the general Fu Guan to rescue the Emperor but failed. He then lived a cloistered life in a cave with his family. Since then, he didn't enter the city for more than 20 years. People praised him as a loyal official to the country and a dutiful son to his mother. Therefore, the rock that he lived in was called Xiao (Filial Piety) Rock.

出世与入世，都在一方岩穴里演绎。

泰宁历史上道教、佛教盛行，而独有的丹霞岩穴，又以其“不涂垦茨而风雨之患除，不凿户牖而日月之光人”的地理特征，成为佛寺、道观的理想福地。

西汉末年，王莽篡汉，人神共愤，梅福弃官而走，来到泰宁栖真岩隐居炼丹；距状元岩约五百米的琵琶岩，是明朝道仙邢德兴修炼之处。现在泰宁不少寺庙是建在丹霞洞穴之间，如甘露岩寺、宝盖岩寺、醴泉岩寺、天台岩寺等。在丰岩，堪为奇观的是，一个岩穴里曾经建了三个寺庙。

泰宁人甚至把寻常日子也过到岩穴里。丹霞岩穴冬暖夏凉，算得上是大自然的厚赠。泰宁杉城镇圣丰岩，一个扁长的岩穴内容纳着一个杨姓村庄。想想人类最初就是从洞穴里走出来的，泰宁的“穴居”也就显得更有韵味了。

◎ 传说当年邹应龙在状元岩苦读几年后从这个峡谷出山，中了状元，当地人把这条峡谷称为“游龙峡”。
Dragon Canyon, where Zou Yinglong came out and ranked 1st in the imperial examination after years of hard working in Zhuangyuan Rock Cave.

◎ 邹应龙的洞里寒窗。
The study cave of Zou Yinglong on a snowy day.

Different lives play out in the caves whether you are part of the religion or not.

Taoism and Buddhism were prevalent in the history of Taining. The unique Danxia caves have become the ideal places for Buddhist temples and Taoist temples with their special geographical features: they are free of wind and rain and there is no need to chisel windows since the light of the sun and moon light the caves.

At the end of the Western Han Dynasty, Wang Mang usurped the power from the emperor, which infuriated both people and the god. Mei Fu abandoned his official position and went to Taining. He lived in solitude in Qizhen Rock and made pills of immortality. The Lute Rock, about 500 meters away from Zhuangyuan Rock, was the place where Taoist priest Xing Dexing did his practice during the Ming Dynasty. Many temples in Taining were built in and amongst Danxia caves, including Ganlu Rock Temple, Baogai Rock Temple, Liquan Rock Temple and Tiantai Rock Temple. At Feng Rock, it is even more amazing that three temples have once been built in one cave.

Taining people even lived their lives in the caves. Danxia caves are warm in winter and cool in summer, which is a wonderful gift from the nature. A narrow but long rock cave located at Shengfeng Rock of Shancheng Town accommodates a village, where the villagers share the same family name—Yang. Considering that the origin of human beings began with cavemen walking out of caves, Taining's "cave-dwelling" appears to be even more charming.

状元岩、李家岩，独孤而处的求学之道

在一年一度的高考来临之前，泰宁以及附近县市有些考生的父母，会带着要高考的孩子，去登一次泰宁的状元岩，希望状元砥砺求学的事迹能让孩子再加把劲，也希望在藏风聚气的宝地沾沾金榜题名的运气。

状元岩在泰宁北郊的杉城镇长兴村，离城区12千米，就在上清溪的隔壁。如果你去上清溪漂流，从下码头上岸后再行不多远，就会到状元岩景区。景区内林木葱郁，属原始次森林，不用说，又是一个让人可以畅快呼吸的氧吧。关键是，这里是泰宁历史上第二位状元邹应龙早年读书求仕的地方。后人咏诗，“此处书声通帝座，当年柳汁染春衫”，这诗写得潇洒，但当时独自躲在这里读书的邹应龙的孤独心境，未必这么浪漫。

如果你有兴致，可以从邹应龙当年走的斗米阶古道上山，想象一个负笈担粮的少年沿着脚下这样曲折盘旋的小小石阶，一步一步默默攀爬上险绝的山崖，潜心隐读，兴许会有更多体会。

在岩穴中苦读之后，邹应龙25岁考上状元，41年后官至副宰相级别，是历史上从泰宁走出的职位最高的官员。只是刚当上副宰相四个月，就有人翻老账弹劾他，他也不辩解，只是长叹一声，当即辞职，皇帝派特使挽留他，他也没有答应。“适谏官有言。即日就道。上再遣中使力谕止，公卒不返。”

邹应龙66岁回到泰宁后，“去家百步，得一丘，曰南谷，有古涧垒岩之胜。日游其间，晨往而夕忘归”，73岁去世。

◎ 邹应龙状元圣像。

A portrait of Zou Yinglong.

Zhuangyuan Rock, Li's Rock, a Lonely Path for Scholars

Before the annual national college entrance examination, parents of some examinees in Taining and the nearby places would take their children to climb Zhuangyuan Rock, hoping that the story of those Zhuangyuan (champion of the imperial examinations) will inspire their children to work harder. Parents also hope their children can have good fortune by visiting this land of blessing.

Zhuangyuan Rock is located in Changxing Village in the northern suburbs of Taining Town, 12 kilometers away from the town center and is just next to Shangqing Stream. If you go water rafting on Shangqing Stream, you will arrive at Zhuangyuan Rock tourist area after a short walk after getting off the raft at the lower dock.

In the tourist area, the trees are lush and luxuriant, forming another oxygen bar for people to breathe freely. More importantly, this was where Zou Yinglong, the second champion of the imperial examination in the history of Taining, studied in his early years. Later, someone wrote a poem about Zhuangyuan Rock, delineating a romantic scene where "the green of the willow trees colour the clothing of the scholars in spring, and the sounds of the readings here pave the way to the imperial capital". For Zou Yinglong who actually stayed here and studied alone, however, perhaps did not have the same romantic feelings.

If you are interested, you may go up the mountain by the ancient trail—the Doumi stone steps that Zou Yinglong had walked on, imagining the scene where a young man carrying his food on the shoulders, climbing the small stone stairs winding around the mountain step by step to the cliffs quietly, living a solitude life and studying hard there. Then you might have some resonance.

Zou Yinglong studied hard in the cave and became the champion of the imperial examination at the age of 25. After 41 years, he was promoted as the deputy prime minister, becoming the highest scholar official in the history of Taining. But his rivals impeached him with old scores only 4 months after. He did not bother to defend himself but sighed and resigned immediately. The emperor sent a special envoy to urge him to stay but he did not agree.

He was 66 years old when he was impeached. After returning to Taining, he lived near the South Valley. The rivers, stones and rocks in the South Valley attracted him so much that he went there every morning and forgot to go home at dusk. This continued until he passed away at the age of 73.

◎ 斗米阶古道。
The Doumi stone stairs that Zou Yinglong had walked on.

◎ 邹应龙读书处。
The cave where Zou Yinglong studied.

邹应龙官至高位，但为官一生似乎并没有什么重大的成就，倒是死后颇为传奇。在客家地区的广东梅州、广西贺州、福建连城和长汀等地，邹氏人家认邹应龙为祖先，并奉其为“广佑圣王”。此信仰甚至传至台湾地区及东南亚等，渐成风气，香火极旺，“广佑圣王”成为台湾地区民众及东南亚华人华侨主要奉祀的俗神之一。

据2007年出版的泰宁文史资料专辑《状元宰辅邹应龙》记载，“宋宝祐六年（1258年），邹应龙逝世后的第十四年，蒙古军大举侵宋，在一次两军激战时，暮云中隐约现出邹应龙的令旗，宋军大受鼓舞，英勇杀敌，战胜蒙古军。宋理宗景定元年（1260年）庚申三月，抗元主帅、京湖制置使马光祖等上表将邹应龙显圣退敌之事奏明朝廷。景定元年（1260年）七月二十八日，理宗特追授邹应龙为‘昭仁显烈威济护国广佑圣王’。”

由此，邹应龙奇妙地完成了从“人间状元”到“天上神仙”的转变。

状元顶，海拔549米，听起来不高，但却是方圆数十里内赤石群的最高峰。站在状元顶，望着一片秀美的山峦起伏地奔天边而去，一样会感慨：“荡胸生曾云，决眦入归鸟。会当凌绝顶，一览众山小。”邹应龙读书的岩穴，就在状元顶下数十米处，外窄内深，长50米，高3米，进深约7米，从地势地形上看，酷似一弯新月。洞内岩壁上凿刻着“状元岩”三个字，洞内有一尊邹应龙的石雕胸像。一座插着线香的神案以及跪拜用的木板垫。除此之外，空荡荡的岩穴留给人们的就只是想象了。

当年邹应龙就坐在这里，青山在眼前，明月白云清风相伴，孤灯把卷苦读。在后来他的女婿赵与筹写的书中，有一段描写邹应龙容貌气质的话：“公风神俊迈，眉目如刻画。半言片简，皆超然有尘外趣，而简静深厚，不求人知。”这气质，也是长时间脱离尘俗、在悬崖绝壁中的修炼才得到的吧。

在山中孤独而处，想得到的不仅是涵养学问，更是修身养性。

◎ 状元顶日出。
The sunrise at Zhuangyuan Peak.

Zou Yinglong ranked a high official in his career, but seemed not have made great achievements. On the contrary, he became a legend after his passing. In the Hakka areas like Meizhou in Guangdong, Hezhou in Guangxi, Liancheng and Changting in Fujian, Zou's family recognized Zou Yinglong as the ancestor and honored him the "Guangyou King". This belief even spread to Taiwan, Southeast Asia and other areas, and gradually became a custom. "Guangyou King" has become one of the main secular gods enshrined by Taiwan people and overseas Chinese in Southeast Asia.

According to the special records of the history and literature of Taining entitled "*Zhuangyuan Prime Minister Zou Yinglong*" published in 2007, "in 1258 which was 14 years after Zou Yinglong's death, the Mongolian army invaded the vast territories of the Song Dynasty. In a fierce battle between the two armies, soldiers of the Song Dynasty had seen Zou Yinglong's flag vaguely appearing through the clouds, which gave them great encouragement. As a result, they fought bravely against the enemy and defeated the Mongolian army. In the third lunar month of 1260, Ma Guangzu, the commander-in-chief of the Song army reported to the court that Zou Yinglong's flag had appeared during the battle and encouraged soldiers to fight hard. On July 28th of the lunar calendar in 1260, Emperor Lizong conferred Zou Yinglong as the "'Guangyou King" to honor him for protecting his country. Thus, Zou Yinglong was transformed from a "champion scholar in the human world" into a "celestial immortal".

Zhuangyuan Peak has an altitude of 549 meters. It may not sound like a tall peak, but it is the highest peak of the red rock group within the larger surrounding area. Standing on top of Zhuangyuan Peak, looking at the beautiful mountains undulating into the horizon, you may have the same feeling with Du Fu when he climbed on Mount Tai: "Clouds rise therefrom and lave my breast; my eyes are strained to see birds fleeting. Try to ascend the mountain's crest: it dwarfs all peaks under our feet. (Translated by Xu Yuanchong)" The rock cave where Zou Yinglong studied was just some dozens of meters below the top of Zhuangyuan Peak. It is narrow at the opening but the inside is deep. It is 50 meters long, 3 meters high and 7 meters deep. Topographically, it looks like a crescent moon. Inside the cave, the wall is engraved with three Chinese characters, "Zhuangyuan Rock". The empty cave only has a stone bust of Zou Yinglong, a case that houses the god sculpture with a line of incense and a wooden mat for worshiping, leaving lots of room for visitors' imagination.

In those days, Zou Yinglong sat here with green mountains in front of his eyes, having the company of the bright moon, the white clouds and breeze of wind, and immersing himself in books. In the book written by his son-in-law Zhao Yuchou, there is a description of Zou Yinglong's appearance and temperament, "He is handsome and has a heroic spirit. His eyebrows are like engravings. His words are always concise, but comprehensive and transcending at the same time. He is simple, quiet with deep thoughts and does not expect others to understand him." Zou may have obtained this temperament from his solitude life in the cliff cave and his detachment from the worldly and the ordinary.

Those who stayed alone in the mountains wanted to get not only the knowledge, but more importantly the self-cultivation.

还有一个离县城更近的是李家岩，最早叫“天台岩”。明朝泰宁籍进士、官至兵部尚书的李春烨，中科举前在这里避世读书，后来李春烨当了朝廷大官，这里就被叫作“李家岩”，再后来李春烨的后人在李家岩内修了家庙，后又改为佛寺，现在叫作“天台岩寺”。

Li's Rock was closer to the county town and had been known as Tiantai Rock earlier. Li Chunye, a Jinshi scholar of Taining who later became the minister of the military department of the Ming Dynasty, had studied here before he succeeded in the imperial examination. Later, after Li Chunye became the important official of the court, the rock was renamed Li's Rock. Li's descendants had first built a temple in the rock cave, which later turned into a Buddhist temple. Now the temple is called Tiantai Rock Temple.

◎ 天台岩寺内景。
Inside of Tiantai Rock Temple.

李家岩在泰宁县城西七八公里的杉城镇际溪村江家坊，与寨下大峡谷相连，从县城出发骑自行车大概也就40分钟。李家岩有丹霞岩槽长达241.3米，已被上海大世界基尼斯确认为“中国最长的丹霞岩槽”。

Li's Rock is located at Jixi Village, 7 to 8 kilometers west of Taining Town. It is connected to Zhaixia Grand Canyon, and takes about 40 minutes from the county town to arrive by bicycle. The Danxia rock groove of Li's Rock is 241.3 meters long, which has been recognized as "the longest Danxia rock groove in China" by Shanghai World Guinness China Records.

◎ 李家岩上中国最长的丹霞岩槽。

Li's Rock has the longest Danxia rock groove in China.

与李家岩同在际溪村境内的还有一个丰岩，也称“瑞丰岩”，由三个相连的岩穴组成，依次为丹霞岩、罗汉岩、丰岩。当地的导游说，南宋抗金名臣、当过宰相的李纲在考中进士前在丰岩读过书。

李纲祖籍邵武，在泰宁隔壁。导游说，李纲的二叔迁居丰岩附近，李纲是到他叔叔家做客时到丰岩读书的。这已无从考证，但可查证的是，李纲的确到过丰岩，但那是他当宰相时被贬海南岛后的事了。在宋建炎四年(1130年)，李纲“自海上来居泰宁”，落脚丹霞岩，拜会名僧宗本禅师，并在这里游山玩水、调理心性，还留下了一篇《瑞光岩丹霞禅院记》描述丰岩风光：“堂殿楼阁，窈窕玲珑，泉石草木，幽奇芳润，叠嶂屏其前，层峦拥其后，山回路转，岩洞乃出，谓造物者融结无意，吾不信也。”后来还有人在岩壁上题写“李忠定公读书处”，至今尚存。

◎ 岩壁上题写有“李忠定公读书处”。
Inscribed on the cliff wall are"Mr Li Zhongding's Study Place".

◎ 瑞丰岩。
Ruifeng Rock (or Feng Rock).

Feng Rock, also known as Ruifeng Rock, is located in the same village as Li's Rock. It consists of three connected caves, namely, Danxia Rock, the Buddha Rock and Feng Rock in consecutive order. According to the local tour guide, Li Gang, the famous anti-Jin officer and the prime minister of the Southern Song Dynasty, had studied at Feng Rock cave before he succeeded in the provincial imperial examination.

Li Gang was born in Shaowu near Taining. The tour guide said that Li studied here when he visited his second uncle who had moved to Feng Rock and lived nearby. It was impossible to verify this information, but what could be found in record was that Li Gang had been to Feng Rock after he was demoted to Hainan Island. In 1130, Li Gang "came to Taining from the sea", and settled in Danxia Rock, met with the famous Zen master, and traveled around to appreciate the scenery, to meditate and calm his mind. He also wrote an article, "*Notes of Danxia Zen Temple*, at *Ruiguang Rock* "depicting the scenery of Feng Rock, "The halls and pavilions are all exquisite; the springs, stones, grasses and tresses are moist, quiet and wonderful. Layer upon layer of mountains like screens stand in front of the caves and there are also mountains surrounded the back of them. Walking through the winding path along mountain ridges, you'll see the caves. I will never believe it is a scene that is inadvertently created." Later, somebody inscribed "Mr Li Zhongding's Study Place" on the cliff wall, which are still there today.

应该是有了邹应龙、李春烨及李纲等人示范在前，在他们身后，越来越多的泰宁学子进山，选择一方属于自己的有缘之地。在泰宁的山山水水间，岩穴所在似乎都传出了一心想考取功名的学子的琅琅读书声。这和千百年来，同样热衷于出现在泰宁岩穴中的道士、僧人一起构成了泰宁最奇特的岩穴人文历史长卷。

19世纪，美国作家梭罗和他笔下的《瓦尔登湖》，隔着遥远的时空呼应着这样与大自然相处的孤独之道："在这美妙的黄昏，我的身心融为一体，大自然的一切尤显得与我相宜。……我独享太阳、月亮和星星，还有我那小小的天地。……喜欢与自然为伍，与我们生命的不竭源泉接近。"

这才是去状元岩、李家岩以及其他岩穴，静处其间后，更值得感悟的吧。

◎ 李家岩禅寺里的消灾泉。
Removing Ill Fortune Spring in Li's Rock Zen Temple.

Inspired by the stories of Zou Yinglong, Li Chunye, Li Gang and others, more and more students in Taining moved into the mountains and found a predestined place that belonged to them. Amongst the mountains and waters of Taining, one can hear the sounds of students reading books— the students who are determined to achieve high scores and good reputation. Over the last hundreds and thousands of years, these students together with the Taoist and Buddhist monks who were also enthusiastic about cave lives delineated the most peculiar history of the cave humanity culture of Taining.

In the 19th century, an American writer Thoreau wrote a book entitled *Walden* that echoed through the distant time and space with such a lonely lifestyle of getting along with the nature: "In this wonderful dusk, my body and mind are integrated into one, and everything seems to be appropriate to me. ... I enjoy the sun, the moon, stars and my little world on my own. ... I like to be with nature and close to the inexhaustible source of our lives."

This is the true meaning that is worth reflecting upon, after staying peacefully and quietly while visiting Zhuangyuan Rock, Li's Rock and other rock caves.

◎ 一览众山小。

Dwarfing all peaks under the feet.

◎ 状元文化公园里反映梅福炼丹的青铜雕塑。
AbronzesculptureofMeiFupracticingalchemyinZhuangyuanCulturalPark.

栖真岩与梅福

公元2年，江西南昌尉梅福，因为不满王莽篡权误国，弃了官，离了家，千里迢迢跑到泰宁，在这里避世修道。

他在距离状元岩半公里的地方，也就是现在上清溪下码头的位置，看中了一个山腰丹霞岩穴，并把这里选作修炼福地，在此结庐炼丹，传说丹成后他羽化成仙，在道教历史上被尊称为“梅福真人”。后来他修炼的这个地方就成了今天的栖真岩。

除了闽越王无诸，最早在泰宁留下遗迹的历史名人就是汉代的梅福了。

Qizhen Rock and Mei Fu

In 2 AD, Mei Fu, the officer of Nanchang in Jiangxi Province, abandoned his official position, left his home and came to Taining after traveling over a long distance. Mei Fu could not stand Mang Wang's usurpation of power and misgovernance and decided to live in solitude and practiced the Taoism.

He found a Danxia cave located halfway up the mountain as a blessed place and stayed there making pills of immortality. The spot is half a kilometer away from Zhuangyuan Rock, which is now the location of the lower dock of Shangqing Stream. After he succeeded in making pills of immortality, he became an immortal himself and was honored as "Meifu Immortal" in the history of Taoism. The place where he did practice was named Qizhen Rock.

In addition to Wuzhu, the King of Minyue Kingdom, Mei Fu from the Han Dynasty was the other historical celebrity who left the traces in Taining in the earliest time.

◎ 栖真岩。
Qizhen Rock where Mei Fu practiced alchemy.

《泰宁县志》说："在长兴保，高二丈许，广五尺余。相传梅福避世炼丹处，今丹炉尚存。中有朝斗石，采药涧。"如今这里剩下的只是一个炼丹炉的基座，还有一个普通的小庙——栖真岩寺。

从山谷边缘的一条石阶小路绕山蜿蜒而上，小路尽头便是寺院的山门。进了山门，左侧是寺院的建筑，右侧则是立陡下去的山谷，很像是山坳。山谷里是一片挺拔高耸的竹林。

偶然看到一个居士在网络上写的在栖真岩寺的生活日记，觉得很有画面感："每天不时有前来拜访师父的出家人和居士，还有山下住的上香者和远来的游客，都静静地来，静静地走。唯有看门的小狗在陌生人到来时偶尔叫几声。""此地空气清新，水清澈且似有一丝甜。吃的是木耳、蘑菇、笋干等山珍，还有门前自种的萝卜、白菜……每天不辍劳作，但不知辛苦，空闲时抓紧拜佛、打坐、诵经，生活充实有序……师父带我去爬山，赏栖霞山貌；带我去山下赶集，买寺院日用所需。我在做这些事的同时，览此地风土人情，好不惬意！"

◎ 梅福的炼丹炉基座。
The base of Mei Fu's alchemy furnace.

◎ 正在打坐的尼姑。
A Buddhist nun sitting in meditation.

According to *Taining County Records,* "(Mei Fu's cave) was located at Changxingbao. It is about 6.6 meters high and 1.67 meters wide. It is said that the alchemy furnace has been preserved. There were also Chaodou Stone and a ravine for washing herbs left in the cave." But now there is only the base of his alchemy furnace and an ordinary temple—Qizhen Rock Temple.

From the edge of the valley, a stone-step path winds its way up the mountain, ending at the gate of the temple. Once entering the gate, the buildings are on the left side, and the steep valley on the right. It is full of bamboo forest inside the valley.

There is a diary written by a lay Buddhist who lives at Qizhen Rock Temple on the Internet. The descriptions in the diary delineate a vivid picture of the lives at the Qizhen Rock: "Every day, there are monks and lay Buddhists who come to visit the Master of the temple, and also worshipers living near the mountain and tourists from afar. They all come and go quietly. Only the puppy who guards the door barks sometimes when strangers arrive." "The air here is fresh; the water is clean and sweet. People eat delicacies of the mountains such as agaric, mushrooms, dried bamboo shoots as well as radishes and cabbages grown in front of the door... We work hard every day and feel tireless. We do worship, zazen meditation and chanting when we have time. We live a fulfilling and orderly life... The Master sometimes takes me to climb the mountain and experience the landscapes of Qixia Mountain; sometimes he takes me to the market at the foot of the mountain to buy daily necessities. While doing all these things, I get to see the local customs, it is really great!"

◎ 栖真岩寺。
Qizhen Rock Temple.

道已不遇，但传说和故事留下了。在泰宁民间，还有许多关于梅福役鬼、捉鬼的故事。“山不在高，有仙则灵”，栖真岩因为梅福会永久地留名于天下。而从栖真岩，我们也窥探到了2000年前泰宁的影子。梅福远在江西都知道跑到泰宁，我们就可以想象，当时泰宁的山水岩穴之中肯定还有更多没有留下姓名的道人的身影。卢真人、邢德兴等许多术士异人之后都在泰宁各地择穴修炼。邢德兴修炼的地方叫琵琶岩，就在离状元岩五百米不到的地方。

中国人内心深处的原始宗教观念充满了对大山敬畏。《礼记·祭法》说：“山林川谷丘陵，能出云、为风雨、见怪物，皆曰神。”尤其是丹霞岩石地貌，以其红色的虚幻庄严和地形的险绝神秘，最早成为道学家构筑神仙世界的蓝本。《山海经》中的“昆仑之丘实惟帝之下都……南望昆仑，其光熊熊，其气魂魂”和《楚辞》中的“仍归人于丹丘兮，留不死之旧乡”都是对丹霞地貌的描述。从汉代起，道教依附传播的名山有很多是丹霞地貌区，如四川的鹤鸣山及青城山、江西的龙虎山及三清山等，泰宁也是丹霞地貌区。

Taoist culture here is no longer flourishing as it once did, but legends and stories have remained. In Taining, there are many stories about Mei Fu enslaving ghosts or catching ghosts. As the saying goes, "A mountain need not be high; it becomes famous so long as there is a deity on it." Qizhen Rock is always remembered because of Mei Fu. At the same time, through Qizhen Rock, we can pry about a view of Taining two thousand years ago. Besides Mei Fu came to Taining from Jiangxi which was far away, many Taoists came to Taining and lived in rock caves. Immortal Lu, Xing Dexing and many other warlocks all chose their rock caves as their place for practicing Taoism. Xing's cave is at Pi-Pa Rock, which is located less than 500 meters away from Zhuangyuan Rock.

The primitive religious ideas in the hearts of Chinese people harbor reverence for mountains. As *The Book of Rites • Sacrifice Law* records, "Mountains, forests, valleys and hills are all deities because they are regarded to be able to manipulate the clouds, form rains and hide monsters." Especially the Danxia rock landscape, with its red illusory solemn and mysterious and dangerous terrains, became a blueprint for Taoist scholars when building an immortal world. *The Classic of Mountains and Seas* and *The Verse of Chu* both include descriptions about Danxia landforms. Since the Han dynasty, many famous mountains where Taoism was spread were Danxia landform areas, such as Mount Heming and Mount Qingcheng in Sichuan, Mount Longhu and Mount Sanqing in Jiangxi and so on. Of course, Taining is also a Danxia landform area.

◎ 栖真岩寺大门。
The gate of Qizhen Rock Temple.

甘露岩寺，距离八百多年的重建

不仅道教，佛教也活跃在泰宁的岩穴中。不同于其他寺庙，岩穴中的寺庙，拥有一个专属的名字——岩寺。

岩寺不以恢宏闻名，但都巧尽地利，灵秀奇巧，与周边环境巧妙地融为一体。

泰宁的岩寺，最最令人惊奇的就是被称为“南方悬空寺”的甘露岩寺。也许在佛教历史上，甘露岩寺并没有什么重要的地位，但作为中国古代营造技艺的杰作，“一柱插地，不假片瓦”的特色已足以让它名闻天下。

奈良被日本国民视为“精神故乡”，是日本古代文化的发祥地之一。到了奈良，不可错过的是奈良公园里的东大寺。东大寺的大佛殿，正面宽57米，深50米，是世界最大的木造古建筑。说起来，它和甘露岩寺还有渊源。甘露岩寺建于南宋时期的1146年，建成若干年后，一个叫重源的日本僧人东渡到中国学习佛法时，曾千里迢迢造访甘露岩寺，对这一建筑之奇叹为观止。不久后他回国，募资重建东大寺大佛殿，在建筑中大量使用“插拱”，就是借鉴甘露岩建筑中的T形头拱设计。

还能证明它金贵的是，它是福建仅存的三座宋代木构建筑（福州华林寺大殿、莆田玄妙观三清殿和泰宁甘露岩寺）之一。

◎ “南方悬空寺”。
The Hanging Temple of the South.

Ganlu Rock Temple and Its Reconstruction after over 800 Years

Like Taoism, Buddhism has also been active among the caves of Taining. Unlike other temples, the temples located in caves have an exclusive name—the Rock Temple.

Rock temples are not known for their magnificence. Instead, they are all delicately beautiful and ingenious, making full use of the geographical location and are integrated with their surroundings wonderfully.

Among all rock temples in Taining, the most famous and amazing one is Ganlu Rock Temple, which is known as "the Hanging Temple of the South". Ganlu Rock Temple did not play an important role in the history of Buddhism, but its skillful design—"the entire temple supported by a pillar plunging into the ground without any tiles" made it well known across the nation. It is considered a masterpiece of the techniques for construction in ancient China.

Nara is regarded as the "spiritual hometown" by Japanese and is one of the birthplaces of ancient Japanese culture. The Todai-ji Temple in Nara Park is a must see when you visit Nara. The Great Buddha Hall of Todai-ji Temple is 57 meters wide in the front and 50 meters deep. It is the world's largest wooden ancient building. You may be surprised at its connection with Ganlu Rock Temple. Ganlu Rock Temple was built in 1146 in the Southern Song Dynasty. A few years after it had been built, a Japanese monk named Zhongyuan came to China to study Buddhism and travel a great distance to visit Ganlu Rock Temple. He was stunned by this architectural wonder. Soon after, he returned to Japan and raised funds to rebuild the Great Buddha Hall of Todai-ji Temple. He used a large number of "arching" in the building, and the idea was borrowed from the T-shaped arch design at Ganlu Rock Temple.

It is precious also because it is one of the only three wooden buildings of the Song Dynasty in Fujian Province, together with the great hall of Hualin Temple in Fuzhou and Taoist Trinity Hall of Xuanmiao Temple in Putian.

大金湖景区建成后，甘露岩寺成了景点之一，参观时需要乘坐游船。乘船登岸，走过莲花桥，甘露岩寺就在眼前不远的赤岩深壑中，途中翠树掩映，蹊径幽邃，10多分钟就可以到达布满苔藓的甘露岩寺山门。导游们总是这样介绍山门的对联：北宋时期，泰宁第一个状元叶祖洽的母亲前来此庙求子，如愿以偿。叶祖洽出生并高中状元，后来还母愿重建了寺庙。这就是一个穿凿附会的美好传说而已，实际上叶祖洽生于1046年，甘露岩寺建于1146年，差了整整100年。

走进山门，抬头就可以看见甘露岩寺建在丹霞岩壁上的一处天然岩穴之中，这个岩穴高80多米，深度和上部宽度均约有30多米，但下部宽度不足10米，呈倒三角形，在地质学上称为“大型单体拱形洞穴”。岩性为红色、紫红色厚层状含钙质的砾岩、砂砾岩以及夹薄层的砂岩、粉砂岩，属湖相沉积。岩层中局部可见鱼尾纹状的交错层理，这是沉积过程中水流方向改变留下的痕迹。

甘露岩寺依着岩穴的自然形态，凭一柱插地撑起四座阁楼——正殿、蜃阁、观音阁、南安阁，阁楼顶部都没有瓦片，下部多以木栈为基础，布局紧凑，错落相接，除了主建筑群之外，还有僧房、生活区，应有尽有，是名副其实的“别有洞天”。

◎ 甘露岩寺山门。

The mountain gate of Ganlu Rock Temple.

◎ 大型单体拱形洞穴。
A large single arched cave.

Ganlu Rock Temple is one of the attractions in Dajin Lake. It is necessary to take a cruise to visit the temple. After getting off the boat and walking through the Lotus Bridge, you can see Ganlu Rock Temple in the deep Danxia rocks not far. Passing the quite path covered with trees, you can reach the mossy gate of Ganlu Rock Temple in about 10 minutes. There is a pair of couplets on the two sides of the mountain gate, and tour guides always tell the story about the couplets this way: during the Northern Song Dynasty, the mother of Ye Zuqia, Taining's first champion of the imperial examination, came to the temple to pray for a son, and her wish came true. Ye Zuqia was born and later became the champion in the imperial examination, thus, he rebuilt the temple to redeem the prayer of his mother. However, this is only a beautiful legend. Ye Zuqia was actually born in 1046 while Ganlu Rock Temple was built in 1146—there was a 100-year gap between.

Walking into the gate and looking up, you can see the temple was built inside a natural cave on the rock. The cave is more than 80 meters high, the depth and upper width are both about 30 meters. The lower width however, is only less than 10 meters. It looks like an inverted triangle and is geologically known as a "large single arched cave". The lithology is a purple and red thick layered calcareous conglomerate, glutenite and thin sandstone and siltstone, which are lacustrine deposits. The crow's feet-like streak layering is partially visible in the rock formation, which is the trace left by the change of the direction of water flow during the process of sedimentation.

Based on the natural form of caves, the four lofts of Ganlu Rock Temple—the main hall, Shen Pavilion, Guanyin Pavilion and Nan'an Pavilion are all supported by one pillar. There are no tiles on the top of the lofts. The lower part is based on the wooden stack and has a compact layout. In addition to the main buildings, there are also dormitories for monks and living areas in the caves. It is indeed a world all on its own.

◎ 甘露岩寺内的僧人。
A monk in Ganlu Rock Temple.

另外，值得一提的是，今天看到的甘露岩寺并不是宋代的原构，而是20世纪80年代重建的。1961年，积蓄了800多年风韵的甘露岩寺被一场大火焚毁殆尽。当时这件事惊动了全国。1961年3月20日，在事故发生一个月后，国务院下文通报，责令全国各级文物保护单位引以为戒，吸取甘露岩寺被焚的教训，切实加强文物保护工作。在金湖还没有诞生、泰宁旅游还没有闻名于世的时候，甘露岩寺的一把让人痛惜的大火，先让泰宁“火”了一把。

令人略微感欣慰的是，重建的甘露岩寺基本忠于原构。

灵山岩穴，钟鼓檀香，这是泰宁地理孕育出的独特风景。不过今日分布在岩穴中的小寺，很多香火冷清，或倾颓无人，或成了居士的住所，还有的成了农家住宅。

It is worth mentioning that Ganlu Rock Temple that we see today is not the original one built in the Song Dynasty. It was rebuilt in the 1980s. In 1961, Ganlu Rock Temple, which had a history of more than 800 years, was burnt down completely. This incident became known across the country. On March 20, 1961, one month after the incident, the State Council issued a document to cultural relics protection office at all levels and instructed them to learn from the incident of Ganlu Rock Temple, and conscientiously strengthen the protection of cultural relics. The deplorable fire of Ganlu Rock Temple made Taining famous even before Dajin Lake had been built and also before Taining becoming a well-known tourist site.

To our relief, the Ganlu Rock Temple that has been rebuilt is basically the same to the original structure.

The mountains and rock caves, the bell, drum and sandalwood, all form a unique scenery born out of Taining's geological features. However, many of the small temples scattered amongst the caves are deserted today. Some of them have fallen into decay, some have become dwellings for laypeople, and some have become the houses of local villagers.

◎ 甘露岩寺木构内景。
The interior wood structure of Ganlu Rock Temple.

圣丰岩，没有屋顶的村庄

“上山不见人，入村不见村，平地起炊烟，忽闻鸡犬声。”在泰宁，利用天然的丹霞岩穴筑房而居，成了独特的丹霞岩穴村落风景，这也是全国独有的天人合一的妙趣。只是随着现代化的足音，大部分岩穴村落早已废弃衰败，墙颓草长。幸而在泰宁县城西北方向六七公里处的圣丰岩，至今还保留着一个杨姓家族村落，村民以险峻的峡谷为寨隘，沿着一条长达100米长的丹霞岩槽建楼，“依穴而居”，也算是人类穴居的“活化石”。

据杨姓族谱记载，居住在圣丰岩的杨姓家族是宋代杨家将客迁入闽的裔孙，入闽期间从福建建阳辗转至南平，后迁回北上，经将乐在明代初期进入泰宁，并在这岩穴里居住了四五百年。20世纪60年代是圣丰岩的鼎盛时期，当时这个自然村算一个生产队，有47户近200个村民，每家每户桌靠桌，凳靠凳，一到吃饭的时候，就像办酒席一样热闹非凡。到2000年前后，住户们为了改善生活条件，开始陆续外迁，只剩几户不愿搬离的和靠山里毛竹讨生活的村民。

◎ 圣丰岩里的古老房屋。
The old houses in Shengfeng Rock groove.

Shengfeng Rock, a Village without Roofs

There's a poem that depicts the village of cave-dwellings, "You cannot find people on the mountain and cannot identify the village after already entered it. But you can see smoke from the chimneys and hear the cocks crowing and dogs barking." In Taining, building houses inside natural Danxia rock caves have become a unique culture. However, with the footsteps of modernization, most of the cave-villages have been abandoned. The walls fall and grass have grown long. Fortunately, there is still a village of Yang's family at Shengfeng Rock, about six kilometres to the northwest of the Taining County. The villagers took advantage of the steep canyon and took it as the village path for building houses along the 100-meter Danxia rock groove. It can be regarded as the "living fossil" of human cave dwelling.

According to the genealogy of Yang's family, the people who lived at Shengfeng Rock was the descendants of the well-known General Yang's Family. They had been to Jianyang, Nanping, Jiangle and finally settled down in Taining, and had lived in the caves for four to five hundred years. The 1960s were the heyday of Shengfeng Rock Village. At that time, the whole village of 47 households with nearly 200 villagers formed a production team. People put all the tables and stools one against another and it was bustling like a feast when it was mealtime. By the year 2000, residents began to move out for better living conditions, only a few who were reluctant to move and those who relied on the bamboos to make a living stayed.

◎ 村落的门牌号。
The house number plate.

◎ 老房屋的土灶。
The old wood-fired oven.

◎ “洞穴飞出的百灵鸟”杨君。
Yang Jun, "the lark that has flown out of a cave".

去圣丰岩，先是一段狭窄的水泥路，最后一段就只是沙土路，路面蜿蜒于赤石峡谷间，只能步行。不久就会看到一道围墙把整个圣丰岩包住，围墙大约两米高，中间有扇大门。进入大门，可见建有房屋的狭长弧形岩穴，长100多米，深20多米，最高处20余米，最矮处不足2米。岩穴上长满了青苔和一些蕨类植物；屋前空地，几畦无人照看的蔬菜兀自生长；池塘里水草一片，舂米的旧家什安静地躺在屋旁。

洞穴离现代文明显得太过遥远。很难想象，上世纪末，从这里走出一位电视明星。小姑娘杨君在这个洞穴里出生成长，后随家庭搬入泰宁县城，初中时参加东南卫视《银河之星大擂台》少儿组比赛，闯过20大关，夺得擂台赛冠军，被称作“洞穴飞出的百灵鸟”。

环视圣丰岩，一种远离尘世之感油然而生。这里静得出奇，空气也格外清新。据说，这里很奇特的一点是没有蚊子，与江西龙虎山的“无蚊村”一样，而且洞穴里冬暖夏凉，通风干爽，村里有通电，山坡上的泉水就是水源。过去，这里的村民寿命都很长，因为这里是一个修身养性、颐养天年的好地方。

Going to Shengfeng Rock, you will pass a narrow cement road at first followed by a section of dirt road, winding through rocks and only allowing people to walk. A wall that is two-meter-high with a gate in the middle surrounds Shengfeng Rock entirely. After entering the gate, you can see the narrow arc-shaped cave with houses in. The cave is more than 100 meters long and more than 20 meters deep. The highest point is more than 20 meters while the lowest is less than 2 meters. Moss and ferns are everywhere. In front of the houses, there are some fields where vegetables grow wildly; water plants grow inside the pond and the tool of the old household for pounding rice lying quietly next to the houses.

Caves seem to be so distant from modern civilization. It is hard to imagine that a TV star came from this cave at the end of last century. A young girl, Yang Jun, was born and raised in the cave, and moved to Taining Town with her family later. When she was in junior high school, she participated in the children's competition titled "the Star of the Galaxy" held by Southeastern Satellite Television of Fujian. She passed 20 rounds successfully, and won the championship of the competition. She has been known as "the lark that has flown out of a cave".

When you look around Shengfeng Rock, feelings of distance from the earthly arise. It is so quiet here, and the air is exceptionally fresh, and it is a "mosquito-free village"! The village has electricity system, and the spring water on the hillside is the water source. Warm in winter and cool in summer, with good ventilation, the cave is a wonderful place for the villagers who used to enjoy longevity.

◎ 沧桑的过往。
The old house records the past of the cave village.

◎ 跳傩舞的泰宁人。（郑友裕 摄）
Taining people performing Nuo Dance.（Courtesy of Zheng Youyu）

远古的姿势——大源村和傩舞

大源村离泰宁县城30多千米，开车约一个小时的路程。村落处在闽赣隘口，古驿道穿村而过，村里人到江西可能比到泰宁县城都近。就是这么一个地方，从最早严、戴两姓开基立业，到现在有1000多年历史了。清朝时，大源村人气最盛，仕宦商旅在驿道上络绎不绝，但这些风光都已消散在历史尘埃里了，剩下的只是依山而建的20多处砖木结构的明清古民居，层层递高。

The Postures of Ancient Times: Dayuan Village and Nuo Dance

Located more than 30 kilometers away from Taining Town. Dayuan Village is situated on the ancient post road between Fujian and Jiangxi provinces. It is probably closer for villagers to get to Jiangxi than to the county centre of Taining Town. Since the founding of the village by the Yan and Dai families, Dayuan Village has had a history of over a thousand years. The village had its heyday in the Qing Dynasty when officers and merchants had traveled along the post road endlessly. However, all these scenes have dissipated in history, and what has remained is more than 20 brick-and-wooden dwellings built along the mountain-sides during the Ming and Qing dynasties.

◎ 傩舞为村民祈福。
Villagers perform Nuo Dance for blessings.

◎ 傩舞面具。
The masks of Nuo Dance.

打开这个1000多年的村落的最佳方式，要在它有傩舞的日子。

在这样的日子，村里傩舞队会先聚齐到严氏祠堂，上身赤膊，下着短裙，每人头戴代表不同神的面具，有弥勒佛、四大天王、祝融火神、风雨雷电水神等。

傩舞队从严氏祠堂出发，沿村巷走遍各个角落。行进中，最前面的两人分别头戴"金童""玉女"面具，手持金元宝、玉如意、蟠桃等道具；随后的两人戴弥勒佛面具，手持木鱼，有节奏地敲击，引领舞队；后面是鼓队，由4到6人组成，手持绘有太极图案的小鼓和鼓槌，敲击、跳跃前进。他们在木鱼声的指挥下，变换队形，进行着不同的动作；其后是尺队，由6或8或12人组成，手持两根长约30厘米的尺板，一样跳跃前进；然后是锣鼓（武乐）队伍和丝竹（文乐）队伍。武乐队伍由两人抬一面大鼓，一人击鼓，两人击锣，两人击大钹，两人击小锣。文乐队伍有笛、唢呐、二胡、京胡等。所奏的乐曲主要有《大开门》《红绣鞋》《夜不宿》等曲牌。再后面是绘有太极图案的一面大旗，旗上写着"风调雨顺""国泰民安"。

整支队伍主要由前头两个拿木鱼的人控制指挥，鼓点的快慢强弱变化均由其协调统一。他们跳跃的动作和武术一样，也有固定的套路和有趣的名字，如"矮子步""踱官步""童子献宝""猫咪洗脸""魁星点斗"等，每个人的行步跨度都很大，动作简单拙朴，但每一下都坚实有力，前后有矩，行走端方，双足跳跃，给人一种扑面而来的狂野和神秘感。

The best way to learn about this 1000-year-old village is to visit it on the days when villagers perform Nuo Dance. On such days, the Nuo dance team of the village will first gather in Yan's Ancestral Hall. The male dancers wear black skirts, each with a mask of different god such as Maitreya, four Heavenly Kings, the Fire Deity, and Deities of Wind, Rain, Thunder and Lightning.

The Nuo dance team will start their performance from Yan's Ancestral Hall and walk around the village. During the march, the two people at the front each wears the mask of the "Jin Tong" and the "Yu Nv" (Jin Tong and Yu Nv are boy and girl attendants of fairies) respectively, while holding the gold ingots, jade Ruyi, flat peaches and other items in their hands. The two people behind wear the masks of Maitreya, hold wooden fish in their hands and beat rhythmically to lead the dance team; behind them is the drum team with four to six people. They all carry small drums with Tai Chi patterns, beating drums with drumsticks and leaping forward. They change the formation and carry out different movements guided by the wooden fish.

The team that follows the drum team is the ruler team, consisting of 6, 8 or 12 people. Everyone leaps forward while holding two 30-centimeter-long rulers. The next is the gong-and-drum team and the Sizhu team. In the gong-and-drum team, two people carry a large drum and a third one beats it, two people beat gongs, two beat cymbals, and the other two beat the small gongs. People in the Sizhu team play different instruments like flute, Suona, Erhu and Jinhu (two-stringed bowed instruments). What they play include the tunes *Opening*, *Red Embroidery Shoes*, and *Sleepless Night*. There is a big flag with a Tai Chi pattern behind the two instrument teams, with some Chinese characters printed on the flag that mean "good weather for the crops" and "may the country be prosperous and people at peace".

The whole team is mainly conducted by the two people at the front with the wooden fish, which means that the speed and strength of the drum beats are coordinated by them. Their leap action is like Chinese martial arts—they also have their fixed patterns and interesting names, such as "dwarf step", "officers' stroll", "a boy presenting a treasure", "cat washing face", "Kuixing Diandou" (According to legend, Kuixing was the deity who dominated the imperial examination. People who wanted to succeed in the examination would worship Kuixing. It is also said that Kuixing had a pen in his hand which is used to point out the names of the successful examinees. Those who dream of the Kuixing will be the lucky ones in the exam) and so on. Everyone's steps are nice and big, their moves are simple, unadorned, but firm and powerful. They keep the right distance as they move forward, walking in squared and proper stride, and jumping on both feet, bringing people a wild and mysterious feeling.

傩，产生于3000多年前的殷商时期，是为了驱邪镇魔而诞生的一种巫舞。南唐时，大源村的严氏老祖宗严续把它从宫廷带回村中，并代代相传至今，成为村里祈福、喜庆的特有仪式。

锣鼓一响，风吹旌幡，傩神走下山间村落逶迤的石阶，跳跃过田塍，穿行过竹林，天地间回荡着一种悠远如上古的粗犷的力量之美，冲击着跑来观看的一拨拨在城市里精致过久的灵魂，让他们寻找到一种质朴刚健的精神放纵，也可以说是一种原始的快乐。

原先大源村的傩舞只在农历正月十五等几个特殊的节庆日子可以看到，如今为了开发旅游，村里的傩舞队也成立了合作社，只要有游客或旅行团预约，他们就会随时聚集表演，成了愉悦旅客的日常节目。

◎ 傩舞双足跳跃。

Jumping on both feet in Nuo Dance.

◎ 傩舞队走村祈福。
The Nuo dance team going through the village to bring blessings.

Nuo was a wizard dance created for exorcising demons and originated in the Yin and Shang dynasties more than 3,000 years ago. During the Southern Tang Dynasty, the ancestors of the Yan family from Dayuan Village brought it back from the palace to the village. The dance has been passed down from generation to generation since then, becoming a unique ceremony for blessing and celebration of the village.

The gongs and drums make sound, the wind blows the flag, and the Deities of Nuo walk down the winding stone steps in the mountain village, jump over the fields and go through the bamboo forests. The beauty of primitive power as distant as in the ancient times resound, leaving an impact on the souls of the visitors who have lived in cities for too long. It has allowed the visitors to enjoy the simple and vigorous spiritual indulgence or a kind of primitive happiness.

Nuo Dance could be seen only during special festivals such as the 15th day of every first lunar month before. Nowadays, the Nuo dance team has established cooperatives in an effort to boost local tourism. The dance team will gather to perform any time booked by tourists or tour groups. The exclusive ceremony of worshiping the deities and earth at restricted times in the past has become a daily program for pleasing tourists nowadays, offering more opportunities for people to appreciate it. This can be considered as the locals keeping up with the times.

慈航回家，白云深处庆云寺

泰宁佛教源远流长，出过不少高僧，如宋朝丹霞岩的宗本禅师、明朝石辋狮子岩的悟空和尚、清朝宝盖岩的煮石道人等，近现代从泰宁走出的慈航法师，又给这个名单添了一笔。

慈航法师俗名艾继荣，福建建宁县人，清光绪二十一年（1895）生，12岁成为孤儿，依舅以活。舅父业裁缝，专门为出家人缝制僧衣。这种因缘让他萌生了出家修道的念头。

1912年，他17岁，在泰宁县峨嵋峰剃度出家，法名慈航。他就读过厦门南普陀寺的闽南佛学院，学承太虚大师，法接圆瑛大师，长期在缅甸、马来西亚、新加坡等地弘法。1948年他受邀到台湾，在台6年，创办佛学院，培养了一大批僧才，声望臻于顶峰。如今名扬世界的星云大师、净良长老等大法师均出自慈航门下。

1950年，慈航法师曾对入室弟子律航说："我离开南洋之前，原打算直接回闽北家乡，买块坟地，然后在附近三个县传布佛法。这次来台湾，只是因利乘便，作一桥梁罢了。"他又曾对弟子星云说："我的祖庭在闽北泰宁，那里山灵水秀，佛缘隆盛。今后，我要带你们到福建家乡看看……"

1954年，慈航圆寂，肉身不坏，成就台湾地区第一尊全身舍利。2007年，其弟子遵照慈航生前"叶落归根、魂归故里"的夙愿，送慈航菩萨圣像回归泰宁峨嵋峰庆元寺祖庭。

据说，慈航在佛学院上课很有特色。他授课时，经常是一上讲台就先唱一首歌，同学们跟着唱："僧青年！僧青年！我们要把新佛教的责任挑上肩！过去的腐败不要去埋怨，未来的建设不要再留连；我们打算怎样去开一条新路线：严持自己的律仪，培养利人的德行，征求应用的知识，实行到民间的宣传。僧青年，僧青年！这就可以作为我们的龟鉴，快快向前，努力，努力！快快向前！"

◎ 慈航法身菩萨回归庆云寺祖庭。

The return of the sacred body of Cihang Buddha to the ancestral Qingyun Temple.

◎ 云雾缭绕的庆云寺。
Qingyun Temple in the fog and mist.

Cihang's Return to Qingyun Temple in the Depths of White Clouds

Buddhism in Taining has a long history with many eminent monks, such as Chan-master Zongben from Danxia Rock in the Song Dynasty, Monk Wukong from Lion Rock in the Ming Dynasty, Taoist priest Zhushi from Baogai Rock in the Qing Dynasty. In modern times, Master Cihang from Taining was one of them.

Master Cihang's secular given name was Ai Jirong. He was born in Jianning County in Fujian Province in 1895 of the Qing Dynasty. Cihang became an orphan at the age of 12 and lived with his uncle since then. His uncle was a tailor that specially sewed monastic garments for monks. Influenced by his uncle's job, Cihang decided to be a monk.

In 1912, at the age of 17, he became a monk at Emei Peak in Taining and was given his religious name Cihang. He studied at Minnan Buddhist College of Nanputuo Temple in Xiamen under the supervision of Master Taixu and Master Yuanying. He had traveled to Burma, Malaysia, Singapore and other places for extensive period of time to spread Buddhism. Cihang was invited to Taiwan in 1948, where he established a Buddhist academy and supervised a large number of monks during the six years of his stay, and his prestige reached the peak. The world-renowned masters such as Venerable Master Hsing Yun and Venerable Ching Liang were both students of Cihang.

In 1950, Master Cihang told his disciple Lvhang, "I had originally planned to directly return to my hometown in northern Fujian before leaving Nanyang, buy a graveyard there and spread Buddhism in three nearby counties for the rest of my life. I wanted to establish a base for the development and growth of Buddhism in the area. Now that I am in Taiwan, I would like to take advantage of the situation and help to build stronger ties." He also told his disciple Hsing Yun, "My ancestral hometown is Taining. It is located in northern Fujian with picturesque scenery and flourishing Buddhism culture. I would like to bring you to my hometown to see Fujian in the future."

Cihang passed away in 1954. His body was preserved well and became the first full-body Buddhist relic in Taiwan. In 2007, his disciples respected his wish to "return home" and sent his statue back to the ancestral home—Qingyuan Temple at Emei Peak in Taining.

It is said that Cihang's classes at the Buddhist academy were very unique. When he taught, he would sing a song at the beginning of the class, and his students would sing along. The song encouraged young monks to shoulder the responsibility of new Buddhism and set up strict rules for themselves, urging them to work hard.

◎ 茶花烂漫映祖庭。

The blossom of camellia flowers at the ancestral temple.

◎ 慈航祖庭。
The ancestral hall of Cihang.

峨嵋峰海拔1714米，是福建省第七高峰，因山势类似四川峨眉山而得名。这里高山草甸辽阔，山茶杜鹃漫山，古松翠柏雄奇。和泰宁大多数建于岩穴中的寺庙不同，位于海拔1600米处的庆云寺始建于宋代，坐落在兜率岭的平展开阔地。这里天气瞬息变幻，有时日出天开，暖阳沐体，有时浓雾涌起，如水墨氤氲，扑于人面，如沐甘霖。庆云寺半隐于雾中，不见周遭重峦叠嶂，苍翠林木，只见山门及大雄宝殿，更显肃穆伟岸，如在仙境。

庆云寺现在的住持是本性法师，也是福州开元寺的方丈，祖籍福建霞浦。他写过一篇关于庆云寺的文章，读来令人神往：

“慈航菩萨纪念堂左侧的小木屋，是寺院的斋堂。有时，村里的牛来此悠闲，用牛角顺便支开木屋柴门，把我们于寺内自种的刚摘取回来的有机青菜甚至豆腐食个干净，然后大摇大摆而去。这些牛，口渴时，还到屋外水龙头下的水罐里喝点水。牛是有主的。偶尔，也有无主的猴子来纪念堂前我们居住的木屋窗外做客。有次，放在窗外的西瓜与饼干，被它们无心地借用了。不过，我们也有担心的，上一段，下了大雪，于雪中，我们竟然发现一路而去的华南虎爪印。谁知道，天籁和咏的夜晚老虎会不会造访我们呢，毕竟伏虎的经验我们尚未有过。

“早些时候，为了建设规划，连续住了几个晚上，夜深人静，月朗星稀，天高星低，空气清新。偶尔的虫鸣鸟叫外，仿佛可听见树与草生长的声音，花开的声音，野果坠地的声音，以及野菇破土的声音。而似乎，动物百灵，却沉沉地酣睡着了，没有了声响。清晰的是，寺院老僧，早起自修，小木鱼的脆声从纪念堂自上而下，悠远而来，漂浮着向空中与山间弥漫而去。这种人、动物、植物、山、水……甚至灵性之物之间的谐和，使我忘却自己还在俗世。如果不是福州的芝山与罗山还有法务需要尽力费时，我真想长栖于此。”

Emei Peak is 1714 meters above the sea level and is the seventh highest peak in Fujian Province. It was given the name because the mountain terrain resembled the well-known Mount Emei in Sichuan Province. There are vast marshy grassland, camellia and azaleas blossom all over the mountains, and ancient pine and cypress trees are magnificent. Unlike most temples that had been built in caves in Taining, Qingyun Temple built in the Song Dynasty was located in the flat and open land on Doushuai Peak with an altitude of 1,600 meters.

The weather here is unpredictable and changes frequently. It is sunny and warm sometimes, feeling like showering with sunshine; other times it is foggy and misty, feeling like bathing in rain. Half of Qingyun Temple hides in the fog, making it difficult to see the surrounding mountains and verdant forests, but only the mountain gate and the main hall of the temple. As a result, the gate and the hall seem to be more solemn and sacred just like a fairyland.

The present abbot of Qingyun Temple is Venerable Master Benxing from Xiapu in Fujian Province. He is also the abbot of Kaiyuan Temple in Fuzhou. He has written a fascinating article about Qingyun Temple:

"The cabin to the left of the Cihang Bodhisattva Memorial Hall is the monastic dining hall. Sometimes, cattle of the village come for leisure, open the wooden door of the cabin with their horns, eat up the freshly-picked organic vegetables that we have planted in the temple, and then swagger away. Those thirsty cattle will also drink some water from the pot under the tap outside. These cattle have their owners, but occasionally, wild monkeys come to visit us at the window of the wooden house in front of the memorial hall. Once, the watermelons and biscuits placed outside the window were taken away by these monkeys. However, we also have concerns. There was a big snowfall some time ago, and we found the footprints of South China tigers in the snow. We are afraid that a South China tiger may visit us one night. After all, we do not have any experience subduing a tiger."

"Earlier, I stayed here for several nights because of the planning for construction. It was quiet at night. The night sky was clear and high with bright stars, the air was so fresh. Occasionally, I could hear the sounds of insects and birds; also, I heard the sounds of growing trees and grass, flowers blossoming, wild fruit falling, and wild mushrooms breaking through the soil. While the animals seemed to fall sound asleep and did not make any noise. The sounds of early morning meditation of the old monks were very clear, the sound of the wooden fish drum came from the memorial hall afar, drifting away in the air and suffused the mountains. This kind of harmony among people, animals, plants, mountains, water…and even spiritual things made me forget that I was still in the real world. If I did not have the work of Zhi Mountain and Luo Mountain in Fuzhou awaiting, I would want to live here for ever."

◎ 杜鹃烂漫峨嵋峰。

Azaleas blossom all over Emei Peak.

自然之歌——留在这里做一位博物家

泰宁大以青年期丹霞地貌为主，兼有花岗岩、火山岩、构造地质地貌等多种地质遗迹，峡谷曲流多姿多彩，是地质科考爱好者的乐园。

类型丰富的地质生态，也相应形成了复杂多样的生态环境，创造了各种植物生长、野兽栖息繁衍的条件，保存了全球性和地方性的珍稀濒危物种。根据《泰宁世界自然遗产地生物多样性研究》一书，泰宁地区有维管束植物共212科645属1412种，脊椎动物36目105科382种，昆虫25目232科1512种，是中国小区域单位面积上野生动植物资源富集区之一。

植物之中，有国家一级保护系列的银杏、南方红豆杉、伯乐树与东方水韭；列入国际自然保护联盟（IUCN）《红色名录》的有铁皮石斛、银钟花、黄山木兰、沉水樟、闽楠、红豆树、伞花木、伯乐树、银杏、长叶榧10种，其中铁皮石斛被列为极危物种；列入濒危野生动植物种国际贸易公约CITES（又称华盛顿公约）附录保护目录的有金毛狗、石仙桃等65种。脊椎动物里，列入IUCN红皮书的有白颈长尾雉、鬣羚、豹、大灵猫等，列入CITES附录保护目录的有游隼、白颈长尾雉、金猫等。

泰宁有特别丰富的兰科植物，在上清溪峡谷中长着心叶球柄兰；在猫儿山长着带唇兰、无柱兰，还有被中医誉为“仙草”的铁皮石斛，以及名字听起来就很美的鹅毛玉凤兰、密花香唇兰、小花蜻蜓兰、小舌唇兰、虾脊兰、伞花石豆兰等，整个泰宁世界自然遗产地的兰科植物多达37属64种。

很多在泰宁出现的植物群落都是珍稀异常的，比如红石沟的闽楠群落、长叶榧树群落、喜树群落；状元岩的黑叶锥群落；甘露岩寺一带季节性洼地中的水毛花群落、睡莲群落；还有猫儿山、状元岩的一些沟谷悬崖上的长叶铁角蕨群落、大花石上莲群落等。

上清溪、寨下大峡谷、九龙潭、大金湖、猫儿山、状元岩、峨嵋峰、九仙崖、金铙山、天成岩等等，这些泰宁已经开发或尚待开发的山水，无一不是认识了解动植物的天堂。暂时撇开人类劳作或人文的印记，泰宁留着一扇更瑰丽多姿的门——与大自然交谈的奥妙之门。

◎ 泰宁长兴村的红豆树群。

Ormosia hosiei trees in Changxing Village of Taining.

The Songs of Nature: Staying Here to Be a Naturalist

The landscape of Taining is mainly composed of youth Danxia landforms as well as granite, volcanic rock, tectonic geological landforms and other geological relics. Together with diverse canyons, rivers and streams, it is a paradise for geological science enthusiasts.

The rich geological and ecological types have provided conditions for the growth of various plants and the reproduction of wild animals, preserving those rare and endangered species on a global and local scale. According to the book *Biodiversity Research of Taining World Natural Heritage Site*, there are a total of 1,412 species, 645 genera and 212 families of plants; 382 species, 105 families, and 36 orders of vertebrate; and 1,512 species, 232 families and 25 orders of insects in Taining area, making it one of the most abundant areas for inhabiting wild flora and fauna in China.

Among the plants, ginkgo, Taxus chinensis, Bretschneidera sinensis and Isoetes orientalis are listed as the national first-class protection series; there are ten species in the *Red List of Threatened Species* of the International Union for Conservation of Nature (IUCN) including Dendrobium officinale, Halesia macgregorii chun, Magnolia cylindrica, Cinnamomum micranthum, Phoebe bournei, Ormosia hosiei, Eurycorymbus cavaleriei, Taxus chinensis, ginkgo, and Torreya jackii. Among them, Dendrobium officinale is listed as an extremely endangered species. There are sixty-five other species, such as Cibotium barometz and Chinese Pholidota Herb, listed in the appendix to *The Convention on International Trade in Endangered Species of Wild Fauna and Flora* (CITES) . Among vertebrates, those included in the *Red List* of IUCN are Syrmaticus ellioti, serow, leopard and zibet, while those listed in the the CITES appendix include peregrine falcon, Syrmaticus ellioti and golden cat.

Taining is particularly rich in orchids. There are as many as 37 genera and 64 species of orchids in the World Heritage area of Taining. For example, there are Mischobulbum cordifolium in the valleys of Shangqing Stream; There are also Tainia dunnii, Amitostigma gracile, and Dendrobium officinale from Maoer (Cat) Mountain, as well as those with beautiful names including Habenaria dentate, Tulotis ussuriensis, Platanthera minor, Calanthe discolor, Bulbophyllum shweliense and the like.

Many plant groups in Taining are rare and unusual, such as the Phoebe bournei group, the Torreya jackii group and the Camptotheca acuminata group at Redstone Ditch; the Castanopsis nigrescens group at Zhuangyuan Rock, the Schoenoplectus mucronatus group and the Nymphaea group at the seasonal low-lying land around Ganlu Rock Temple; the Asplenium prolongatum group on the gully cliffs of Maoer Mountain and Zhuangyuan Rock, and the lotus group on Dahua Stone.

The developed or untapped landscapes in Taining, such as Shangqing Stream, Zhaixia Grand Canyon, Jiulong Pond, Dajin Lake, Maoer Mountain, Zhuangyuan Rock, Emei Peak, Baxianya Peaks, Jinnao Mountain, Tiancheng Rock and so on, are all heaven for naturalists who want to learn about plants and animals. Setting aside the imprints of the works of the humans or the marks of the humanity temporarily, Taining opens an even more magnificent door—the door of wonder for having a dialogue with the nature.

◎ 泰宁大田乡的千年银杏树。
A 1000-year-old ginkgo in Datian Township of Taining.

峨嵋峰上东海洋

地球生命的演变，经历了从海洋到陆地的过程，距今5亿年前，植物率先登陆，改变了地球大气环境，使氧气成为地球大气的主要成分之一。植物的茂盛生长，为动物登陆提供了条件。蕨类植物是植物界中最早登上陆地的种群，至今留有12000余种，仍在不断发展演化。蕨类植物没有种子，靠根部的孢子结合进行“传宗接代”，其传播方式主要依靠水流和鸟类。

在中国南部的一些湖沼湿地，生长着5种形似普通韭菜的蕨类植物——中华水韭、云贵水韭、台湾水韭、东方水韭和高寒水韭。它们是我国特有物种，在地球上生活已经超过3亿年，全部被列为国家一级保护植物。其中，东方水韭目前在全世界数量只有几百丛，属于世界级珍稀濒危植物，由武汉大学生命科学院专家于2002年在浙江松阳县首次发现。2012年，厦门大学研究人员在泰宁峨嵋峰自然保护区“东海洋”沼泽湿地中，又发现了两个东方水韭的自然居群，其中一个居群有200余株，大部分为幼苗；另一居群约有110株两年以上的植株，殊为惊奇。

峨嵋峰发现东方水韭的地方，在距离主峰三四千米处的一个古老的死火山口，现在成了高山中的沼泽湿地，海拔在1500米左右，面积有140公顷，各种飞禽走兽昆虫惬意地栖息其间。近看高山沼泽，其实就是一个水塘连着一个水塘，把山地连成了一片水的世界。水都无比清浅，水中活动着些看得见但叫不出名字的小生物。蹲下身子，用手轻轻在水上拂过，千万要小心温柔，因为侧边的那些不起眼的水草也许就是世界珍稀濒危物种——东方水韭。

◎ 东方水韭生长地。

The place where the Isoetes orientalis grow.

◎ 东方水韭。
The Isoetes orientalis.

The "East Ocean" Wetland at Emei Peak

The evolution of lives on the earth has gone through the process from the ocean to land. Plants took the lead in landing 500 million years ago, changing the atmosphere of the earth and making oxygen one of the main components of the earth's atmosphere. The growth and the flourishing of plants provided the conditions for the birth of animals. The ferns were the earliest species of plants that had grown on the earth, and there were still more than 12,000 species of ferns that remained. They had kept evolving until now. Ferns, which have no seeds, rely on the combination of spores in the roots to reproduce. They depend mainly on water flow and birds for transmission.

There are five species of ferns that resemble chives at some lakes and wetlands in southern China — Isoetes sinensis, Isoetes yunguiensis, Isoees Taiwnaensis Devol, Isoetes orientalis, and Isoetes hypsophila. They are endemic to China and have been living on the earth for more than 300 million years, all of which are listed as national first-class protected plants. Among them, the Isoetes orientalis has a total of a few hundred clusters, which makes it a world-class endangered specie. It was first discovered in Songyang County of Zhejiang Province in 2002 by experts from the faculty of life sciences of Wuhan University. In 2012, researchers from Xiamen University discovered two more natural groups of Isoetes orientalis in the "East Ocean" wetland of Emei Peak Nature Reserve in Taining, one of which had more than 200 plants, mainly composed of seedlings. There were 110 plants older than two years in age in the other group, which was a big surprise.

The place where the Isoetes orientalis was found around Emei Peak was an ancient extinct crater located at approximately three or four kilometres away from the main peak. It is now a swamp wetland in mountains with an altitude of 1,500 meters and an area of 140 hectares, housing a variety of birds, animals and insects. Upon close observation, it is a pond connected to another, connecting the mountain land and turn them into a water world.The water is extremely clear and shallow, and there are some small creatures in the water that can be seen but cannot be identified by names.You should be delicate when squatting down and gently slap on the water by hand, because the inconspicuous grass on the side may be the rare and endangered species in the world—the Isoetes orientalis.

◎ 东方水韭。

The Isoetes orientalis.

峨嵋峰藏着的惊喜不止于此。峨嵋峰位于泰宁县北部的新桥乡境内，距县城24千米，这里四季皆美，春花烂漫、夏风习习、秋意满山、冬雪皑皑，尤其夏天时是有名的避暑胜地。峨嵋峰国家自然保护区是泰宁县境内植物保存最为完好的地带。

2013年，台湾荒野国际保护协会在泰宁峨嵋峰自然保护区考察，发现了台湾镰翅绿尺蛾和砚蝶凤蛾。镰翅绿尺蛾原本被认为是台湾独有，这次是在大陆地区首次发现，而砚蝶凤蛾也是在福建第一次被发现。

沿山而上，在海拔500—850米的地方遇到的主要是杉木、茶果人工林、毛竹、丝栗栲、甜槠及常绿阔叶林，山涧之间彼岸花纤细的花梗把数朵排成伞形的红花高高托起；爬到海拔850—1200米处，你还会看到马尾松、槠栲及毛竹等；在海拔1200—1550米处，主要植被就变成天然黄山松、马尾松、柳杉等；海拔1550米以上，就有了矮林及灌木地带，还有山地苔藓矮曲林。在这个海拔，还能看到台湾野核桃林、半枫荷、亮叶水青冈、红花油茶树等。红花油茶树花期很长，从每年的11月到第二年的4月都有开花，火红耀眼。再往高处，就是高山草甸沿着起伏的山岭生长，极目尽是一片柔软的绿色。

泰宁峨嵋峰的鸟类非常丰富，特别是在春季，会吸引众多摄影师组团来拍鸟。在摄影论坛上看到一个厦门摄影师的留言：“峨嵋峰自然保护区里，长期生活着福建观鸟人心目中的三大神鸟：黄腹角雉、白颈长尾雉和勺鸡……三年前一次偶然的机会去了一趟峨嵋峰，发现自己喜欢上这座山了。从此一发不可收拾，三年间去了八趟，每趟来回一千多千米！不仅仅是富贵态十足的黄腹角雉、少数民族王子般的白颈长尾雉和气势昂扬的竹林王子勺鸡让我情有独钟，就是那悠闲漫步的白鹇，那嘤嗡作响的各种小鸟，那竹林松风，都会使我流连忘返。”

◎ 峨嵋峰春色。
Emei Peak in spring.

There are many more surprises that are hidden at Emei Peak. Emei Peak is located in Xinqiao Township in the north of Taining County, which is 24 kilometers away from the county town. It has different charms in all four seasons. It has romantic flowers in spring, breezy winds in summer, colorful mountains in autumn and white snow in winter. It is a famous summer resort. Emei National Nature Reserve is the most well-preserved area for plants in Taining County.

In 2013, experts from Taiwan Wilderness Society had a field trip in Emei Peak Nature Reserve in Taining. They found the Tanaorhinus formosanus and Yandie Epicopeiidae during the trip. Before this discovery, Tanaorhinus formosanus was thought to be unique to Taiwan, and this was the first time that it had been found in the mainland of China. Meanwhile, Yandie Epicopeiidae was also discovered in Fujian for the first time.

Climbing up the mountain, you will see Chinese firs, tea plantation, bamboo, Castanopsis fargesii, Castanopsis eyrie and evergreen broad-leaf forest grow at the altitude of 500 to 850 meters. The slender stalks of equinox flower hold up several umbrella-shaped flowers by the mountain streams. When the altitude reaches 850 to 1,200 meters, there grow Masson pines mixed with bamboos. When climbing up to 1200 to 1550 meters above sea level, you will see the main forests including Huangshan pine, Masson pine, and Cryptomeria, etc. At an altitude of above 1550 meters, there are low forest and bushes, and mountain moss low forest. At this height, you can also see Taiwan wild walnut forest, Pterospermum heterophyllum, Fagus Lucida, safflower camellia and the like. Safflower camellia has a long period of blossoming, which goes from November until April of the following year. The flowers are red and ablaze. When reaching higher, you will be delighted at the alpine meadows growing along the undulating hills as far as the eyes can see.

Emei Peak is rich in species of birds particularly in spring, when photographers come in groups for the birds. I once read a blog at an online forum for photography that had been written by a photographer from Xiamen: "In Emei Peak Nature Reserve, there are three kinds of birds that are very special to bird-watchers in Fujian: Cabot's tragopans, Elliot's pheasants and pukras... Three years ago, I visited Emei Peak and found myself really fascinated by the mountain. Since then, I had been there eight times in three years, each roundtrip over 1,000 kilometers. I felt passionate not only for the buxom Cabot's tragopan, minority-prince-like Elliot's pheasants and the high-spirited prince of the bamboo forest, pukras, but also the leisurely-wandering Silver pheasants, the humming small birds and the breeze of the bamboo forest all made me linger."

◎ 黄腹角雉。
A Cabot's tragopan.

◎ 白颈长尾雉。
An Elliot's pheasant.

◎ 勺鸡。
A pukras.

红石沟，投入原始的呼吸

泰宁丹霞有深切峡谷470多条，纵横之间，构成了密集的网状谷地，随便进入一处，都是植物王国、昆虫乐园、动物世界、鸟的天堂……红石沟与上清溪相邻，与九龙潭隔路相望，距离县城15千米。由于物种的多样性和独特性，红石沟被国际自然保护联盟的专家推荐为申报世界遗产科考线路之一。

初入沟谷，两岩低峙，越往前行，峡谷越开阔，崖壁上土壤少得可怜，却从峡谷沟底到悬崖顶部依次分布着卷柏、喜树、闽楠、黄花菜、长叶榧、尖叶栎等多样植物群落，其中国家二级保护植物喜树、闽楠、长叶榧三大群落数量都在1000株以上，是其他丹霞申遗地所没有的特有种群。

红石沟长约三四千米，出口和入口距离不到200米，在里面绕上一圈，恰似进入一个天成的自然博物馆。

◎ 进入红石沟。
Entering Redstone Ditch.

◎ 红石沟郁郁葱葱的树林。
The dense forest at Redstone Ditch.

Coming to Redstone Ditch, Immersing in Primitive Breath of Nature

There are more than 470 deep cutting Danxian gorges in Taining, which form dense net-like valleys between the vertical and horizontal lines. These valleys are plant kingdoms, insect paradises, animal worlds, bird heavens... Redstone Ditch is adjacent to Shangqing Stream and faces Jiulong Pond across the road, 15 kilometers away from the county town. Due to the species diversity and uniqueness, Redstone Ditch has been recommended by experts from IUCN as one of the World Heritage scientific research routes.

The cliffs on the two sides of the entrance of the valley are low, but the further you go, the wider the canyon is. The soil on the cliffs is negligible, but from the bottom to the top of the cliffs grow Selaginella, Camptotheca acuminata, Phoebe bournei, Daylily, Torreya jackii, Quercus oxyphylla and so on. Among them, the amount of the three largest national second-class protected plant groups including Camptotheca acuminata, Phoebe bournei and Torreya jackii grow, with the total numbers of more than 1,000 plants each. These plants are unique to Redstone Ditch and cannot be found at any other world heritage Danxia site.

Redstone Ditch is about three to four kilometers long. The distance between the exit and the entrance is less than 200 meters. Walking around it is just like visiting a natural museum.

卷柏，又名“九死还魂草”，因耐旱力极强，在经历长期干旱后根系只要在水中一浸泡后就又可舒展，故而得名。根能自行从土壤中分离，蜷缩似拳状，随风移动，遇水就能再活过来，根重新钻到土壤里寻找水分。

长叶榧属红豆杉科，是新生代第三纪残存的孑遗裸子植物，距今约两亿年左右，是中国特有的珍稀树种。泰宁又是目前在中国发现的长叶榧分布最多、范围最广的区域。除了红石沟，寨下大峡谷、上清溪、天成岩也有发现长叶榧。

◎ 长叶榧。
A Torreya jackii.

Selaginella, also known as "nine-death-and-resurrection grass", is extremely resistant to drought. After a long period of drought, the root system can stretch as long as it is soaked in water. That is how it has gotten the name. Its roots can be separated from the soil by itself, curl up in a fist shape and move with the wind. It becomes alive again as soon as it encounters water, and the root can get into the soil to search for water.

Torreya jackii belongs to Taxaceae, which is the remains of Cenzoic tertiary gymnosperms from about 200 million years ago. It is a valuable and rare variety of trees endemic to China. Taining is the area with the largest distribution and the widest range of Torreya jackii in China. In addition to Redstone Ditch, Torreya jackii can also be found at Zhaixia Grand Canyon, Shangqing Stream and Tiancheng Rock.

闽楠，一直被视为最理想、最珍贵、最高贵的建筑用材，武夷山白岩的楠木船棺，距今已有3000多年历史，仍保存完好。

喜树是中国特有的一种高大的落叶乔木，喜欢生长在潮湿的沟边或溪边，在李家岩至寨下的沟谷中能见到。沟谷中还有高大笔直的栲树、苦槠树，与附着的藤蔓竞争向上。

沟里各种可用作中草药的植物不计其数，红石沟简直可以被称作“药谷”。金针花、龙须草，也一丛一丛附壁生长；斛蕨成排，像鹅毛笔；檵木，枝条斜逸，细叶如织，在清明节前后开花，花可以煮凉茶，叶子也是药材，捣烂了可以止血；旋蒴苣苔，可以治疗中耳炎，跌打损伤；大花石上莲，生在林下岩石上，鲜用或晒干都可以治肺热咳嗽……

◎ 闽楠树群。

The Phoebe bournei group.

Phoebe bournei has always been regarded as the most ideal, most precious and noblest material for construction. The Nanmu boat-shaped coffins made of this material at the White Rock in Mount Wuyi has had a history of more than 3,000 years and are still well preserved.

Camptotheca acuminata is a genus of tall deciduous trees unique to China. They often grow at the edge of damp gullies or streams, and can be seen in gullies around Li's Rock and Zhaixia Grand Canyon. Besides, tall and straight katus and Castanopsis sclerophylla compete with the attached vines to grow higher.

In the valley of Redstone Ditch, there is a variety of plants that can be used as Chinese herbal medicine, which makes it a "Medicine Valley". For instance, the datlily and Chinese alpine rush attach to the wall and grow clump by clump. The drynaria fortunei grows in a row, which looks like a row of quill pens. Loropetalum chinense with slanting branches blossom before or after Tomb-Sweeping Day. The flowers can be used to cook herbal tea, and its leaves are also medicinal materials, which can stop bleeding after being smashed. Boea hygrometrica can treat tympanitis and traumatic injury. The oreocharis maximowiczii that grow on the rocks under trees can cure cough and lung heat whether they are fresh or dry.

◎ 苦槠树。
A Castanopsis sclerophylla.

◎ 白鹇。

A silver pheasant.

沟谷中百鸟争鸣，珍稀的白鹇在谷中悠闲觅食，黑耳鸢、蛇雕、林雕、鹰雕、白腹隼雕或凌空盘旋，或在峭壁岩洞衔枝筑巢。听说还有熊，会游泳上树。

坐在红石沟斜谷的木梯上，看满山漫谷的草木竞相生长，虫鸣声响起了，世界安静下来，听到自己呼吸着湿润洁净的空气，慢慢全身放松，思绪开始轻盈透明，感觉自己也和它们一样，微小但真实。

现代人拥有很多社会知识，使用着高科技的工具，但朴素地感知我们赖以生存的自然世界的能力却越来越退化。

博物，是一种值得拥有的获得朴素快乐的能力。

著名博物学家E.O.威尔逊说：“世界的每一个角落都有无限的活力，等着人们去探索，哪怕只有片刻。至于那些所谓的‘现代科技的奇迹’，我要提醒读者：即使是路边的杂草或者池塘里的原生生物，也远比人类发明的任何装置要复杂难解得多。”

把自己放入大自然，亲近认识大自然，收获的不仅是知识上的满足，“它同时提供情怀、世界观和人生观”，能让你以一种不同的态度和眼光看世界，可能让你更容易拥有全球视角，知道和谐共生的重要性。

There are hundreds of various kinds of birds in the valley having an ongoing "singing competition". The precious silver pheasants leisurely look for food; black-eared kites, crested serpent eagles, black eagles, hawk eagles, and bonelli's hawk eagles hover in the sky or nest in a cliff cave. It is said that there are also bears and they can swim and climb up trees.

Sitting on the wooden ladder in the valley of Redstone Ditch, looking at the grass growing while listening to the chirping of the insects, you will feel your inner world quiets down. You could hear your own breathing, in and out the clear and moist air. The body gradually relaxes and the mind is clear and calm. You feel just like them, tiny but real.

People in modern societies have plenty of social knowledge and use high-tech tools. However, their ability to simply feel and learn the natural world that we live in is increasingly degraded. Having broad knowledge about things is an ability for gaining simple happiness that is worth having.

The well-known naturalist E.O. Wilson said, "Every corner of the world has infinite vitality, waiting for people to explore, even if it is only for a moment. As for the so-called 'miracles of modern technology', I want to remind readers that even the weeds of a roadside or the protists in the pond are far more complicated and incomprehensible than any device invented by humans."

Immersing yourself in nature, getting close and getting to know it will not only bring you the fulfillment of having the knowledge, "but also provides you with feelings, world view and outlook of life". This allows you to see the world with a different attitude and vision. You may gain a global perspective more easily and understand the importance of harmonious symbiosis of man and nature.

◎ 南方铁杉。
A Tsuga chinensis.

◎ 南方红豆杉。
A Taxus chinensis.

03

旅行规划

经典游一日线路

■ 线路一

大金湖（坐游船，参观大赤壁、甘露岩寺、幽谷迷津等）——**泰宁古城区**（主要参观尚书第、世德堂）——**梅林戏展示中心**（听梅林戏）——**尚书街**（品尝小吃擂茶、游浆豆腐、暖菇包等）

■ 线路二

泰宁地质博物馆（有泰宁所有景点的缩微模型和文字介绍）——**寨下大峡谷**（泰宁修建的第一条地质科考路线，被联合国专家称为“世界地质公园的榜样景区）——**李家岩**（明兵部尚书李春烨读书处，李家岩丹霞岩槽据称是“中国最长的丹霞岩槽”）——**丹霞岩**（南宋宰相李纲读书处）

■ 线路三

上清溪（坐竹筏，游览文人墨客心中颜值最高之地）——**栖真岩**（梅福真人的修炼处）——**状元岩**（南宋状元邹应龙读书处）——**九龙潭**（坐竹筏，游览组成密度和复杂度最高的中国丹霞峡谷群）

■ 线路四

明清园（有民间“中国木雕艺术博物院”之誉）——**大源村**（看古民居、观傩舞，观傩舞需提前与表演队联系预定）

深度游主题线路

■ **穿越泰宁历史之旅（一日）**

金饶山（汉闽越国的无诸王游猎处，泰宁世界地质公园四大地质景观园区之一）——**宝盖岩**（唐末五代十国时期开泰公邹勇夫崖葬处，还有千年历史的宝盖岩寺）——**石辋大峡谷和南石寨**（泰宁世界地质公园四大地质景观园区之一，南石寨是明清之际古战场）——**泰宁城区红军街**（红军革命遗址遗迹）

■ **大自然博物之旅（一日）**

猫儿山国家森林公园（从另一个角度欣赏大金湖，还可参观黄石寨、虎头寨）——**峨嵋峰国家自然保护区**（峨嵋峰是福建省第七高峰，可观赏到大量珍稀动植物；还可参观庆云寺）——**红石沟**（由于物种的多样性和独特性，被国际自然保护联盟专家推荐为申报世界遗产科考线路之一）

■ **户外徒步之旅（一日）**

八仙崖（泰宁世界地质公园四大地质景观园区之一，中国东南海拔最高的丹霞山峰）—— **石辋大峡谷**（泰宁世界地质公园四大地质景观园区之一）

Tours Recommended

Classical Day-Trip Tours

Option 1—1 Day

- Dajin Lake—Taking a cruise to visit the Grand Red Cliff, Ganlu Rock Temple, Valley Maze etc.
- Taining Ancient Town—Mainly to visit Shangshu Mansion and Shide Hall.
- Meilin Play Exhibition Center—To watch Meilin Play.
- Shangshu Street—To taste pounded tea, "swimming" tofu, mushroom dumplings and so on.

Option 2—1 Day

- Taining Geology Museum—With miniature models of all attractions in Taining accompanied with detailed introductory text.
- Zhaixia Grand Canyon—Taining's first geological research route, which is recognized by experts from the United Nations as "a model site for world geoparks".
- Li's Rock—The study site of Li Chunye, the minister of the military department of the Ming Dynasty; the Danxia rock groove of Li's Rock is said to be "the longest Danxia rock groove in China".
- Danxia Rock—The study site of Li Gang, the prime minister of the Southern Song Dynasty.

Option 3—1 Day

- Shangqing Stream—Drafting on the most beautiful place in the hearts of the literati.
- Qizhen Rock —The place where Master Mei Fu practiced alchemy.
- Zhuangyuan Rock—The study site of Zou Yinglong, the champion of an imperial examination in the Southern Song Dynasty.
- Jiulong Pond—Taking a bamboo raft to visit the Danxia canyon group with the highest density and complexity in China.

Option 4—1 Day

- The Mingqing Palace—Known as the folk "Woodcarving Art Museum of China".
- Dayuan Village—To visit the historical dwellings, watch Nuo Dance performance, appointments are required in advance.

In-Depth Theme Tours

■ Option 1—1 Day

A journey through the history of Taining

- Jinnao Mountain—The hunting site of Wuzhu, the king of Minyue Kingdom; one of the four geological landscape parks of Taining Global Geopark.
- Baogai Rock—Cliff burial site of Zou Yongfu of the late Tang Dynasty; Baogai Rock Temple, a temple with a history of over a thousand years.
- Shiwang Grand Canyon and the South-Stone Village—The canyon is one of the four geological landscape parks in Taining Global Geopark; the South-Stone Village is a battlefield during the Ming and Qing dynasties.
- Red Army Street in Taining Town—The site and relic of the Red Army revolutionaries.

■ Option 2—1 Day

A nature tour

- Maoer Mountain National Forest Park—To appreciate Dajin Lake from another angel; you could also visit Yellow Stone Fortress and Tiger Head Fortress.
- Emei Peak National Nature Reserve—The seventh highest peak of Fujian Province and a paradise for some rare plants and animals; you could also visit Qingyun Temple.
- Redstone Ditch—Due to the diversity and the uniqueness of the species, Redstone Ditch has been recommended by experts from the IUCN as one of the World Heritage scientific research routes.

■ Option 3—1 Day

An outdoor hiking tour

- Baxianya, or Eight-Immortal Peaks —One of the four geological landscape sites of Taining Global Geopark; it is also the Danxia Mountain with the highest altitude in southeastern China.
- Shiwang Grand Canyon—One of the four geological landscape sites of Taining Global Geopark.